THE ORIGINS OF LIFE

*From the birth of life
to the origin of language*

JOHN MAYNARD SMITH, FRS

and

EÖRS SZATHMÁRY

OXFORD
UNIVERSITY PRESS

OXFORD

UNIVERSITY PRESS

Great Clarendon Street, Oxford OX2 6DP

Oxford University Press is a department of the University of Oxford
and furthers the University's aim of excellence in research, scholarship,
and education by publishing worldwide in

Oxford New York

Athens Auckland Bangkok Bogotá Buenos Aires Calcutta
Cape Town Chennai Dar es Salaam Delhi Florence Hong Kong Istanbul
Karachi Kuala Lumpur Madrid Melbourne Mexico City Mumbai
Nairobi Paris São Paulo Singapore Taipei Tokyo Toronto Warsaw

with associated companies in Berlin Ibadan

Oxford is a registered trade mark of Oxford University Press
in the UK and in certain other countries

Published in the United States
by Oxford University Press Inc., New York

© John Maynard Smith and Eörs Szathmáry, 1999

The moral rights of the author have been asserted

Database right Oxford University Press (maker)

First published 1999

First issued as an Oxford University Press paperback 2000

British Library Cataloguing in Publication Data
Data Available

Library of Congress Cataloging in Publication Data
Data Available

ISBN 0-19-286209-X

3 5 7 9 10 8 6 4 2

Typeset by
Footnote Graphics, Warminster, Wilts

Printed in Great Britain by
Cox & Wyman Ltd,
Reading, Berkshire

THE ORIGINS OF LIFE

John Maynard Smith is Emeritus Professor of Biology at the University of Sussex.

Eörs Szathmáry is at the Institute for Advanced Study, Budapest.

PREFACE

In 1995, we published a book, *The major transitions in evolution*, which presented a novel picture of evolution. Our basic idea was that evolution depends on changes in the information that is passed between generations, and that there have been 'major transitions' in the way that information is stored and transmitted, starting with the origin of the first replicating molecules and ending with the origin of language. Our aim was to explain these transitions in consistently Darwinian terms.

That book was aimed at professional biologists, and assumes considerable prior knowledge. In this present book, we present the same ideas to a general readership. Although we have assumed little prior knowledge of biology, we fear that it will not be an easy read. It contains a lot of facts, and a lot of new ideas. But we believe that anyone willing to make the effort will end up understanding many of the most fundamental ideas in biology.

We have received a lot of help along the way. E.S. is grateful to the Collegium Budapest and its former Rector, Lajos Vékás, for their support while the book was being written, and to the many visitors to the institute, especially in the years 1994–5 and 1996–7, who contributed ideas. J.M.S. owes a similar debt to the School of Biological Sciences at the University of Sussex, and to the Swan at Falmer. We owe much to our editor, Michael Rodgers of OUP, in particular for telling us when we had failed to make an idea clear, and in arranging for comments on the manuscript by Helena Cronin, Mark Ridley, and Colin Tudge. It is hard to imagine a trio better able to advise on a project like this one: we are grateful to all of them. Finally, our thanks are due to our copy editor Sally Bunney for the meticulous way in which she corrected our errors and infelicities. With such helpers, there can be little excuse if we have failed to write the book we were aiming at.

November   J.M.S.
1998   E.S.

CONTENTS

Preface *vii*

1 Life and information *1*

2 The major transitions *15*

3 From chemistry to heredity *31*

4 From the RNA world to the modern world *37*

5 From heredity to simple cells *47*

6 The origin of eukaryotic cells *59*

7 The origin of sex *79*

8 Genetic conflict *95*

9 Living together *101*

10 The evolution of many-celled organisms *109*

11 Animal societies *125*

12 From animal societies to human societies *137*

13 The origin of language *149*

Further reading *171*

Glossary *173*

Index *177*

CHAPTER 1

LIFE AND INFORMATION

Organisms are incredibly complex. The more we know about them—their biochemistry, their anatomy, their behaviour—the more astonishing are the detailed adaptations that we discover. How could all this complexity have arisen? We can all understand that, if a farmer breeds from the cows that give most milk, the milk yield of the cows will in time increase. This depends on the fact that offspring resemble their parents—that is, on heredity—but this does not at first sight present any great difficulty: we are familiar in our daily lives with the fact that children resemble their parents. Darwin's theory of evolution by natural selection amounts to no more than the idea that, in nature, those individuals best able to survive and reproduce will transmit the characteristics that enabled them to do so to their children, leading to the evolution of traits beneficial to the organism itself, rather than to the breeder. It is, perhaps, the one profound idea in science that we can all readily understand.

Although Darwin's idea is simple—perhaps because it *is* so simple—it is hard to believe that it can really explain the complexity we see around us. We may be able to breed cows that give more milk, but we could not breed pigs that fly, or horses that talk: there would be no promising variants that we could select and breed from. Where does the variation come from that has made possible the evolution of ever increasing complexity? Biology textbooks are liable to say that mutations—that is, new heritable variants—are random. The statement is near enough true, although 'random' is a notoriously difficult word to define: it would be better to say that, in general, new mutations are more likely to be harmful to survival than adaptive. Can it really be true that mutations that in their origin are non-adaptive led to the evolution of the wonderfully adapted organisms we see around us? This book is an attempt to answer that question. In trying to answer it, we shall be led to review a large part of modern biology.

The crucial step is to understand the mechanism of heredity, because the whole process of evolution by natural selection depends on it: if children did not resemble their parents, the Darwinian mechanism would not work. So, how can structures be replicated? The simplest way is by 'template reproduction'. If you wished to copy a metal statue, you would first make a clay mould of the statue and then pour molten metal into the mould, thus obtaining a copy of the original. This is an effective way of copying a surface, but would not work if the object to be copied resembled a living organism with a complex internal structure, down to the level of cells and molecules. In fact, when a complex organism reproduces, it first produces a simple egg, which then develops into an adult. The egg contains complex molecules, but no morphological structures similar to those of the adult: there is not a homunculus in the sperm.

How is it, then, that an egg develops into a mouse, or an elephant, or a fruit fly, according to the species that produced it? The short answer is that each egg contains, in its genes, a set of instructions for making the appropriate adult. Of course, the egg must be in a suitable environment, and there are structures in the egg needed to interpret the genetic instructions, but it is the information contained in the genes that specifies the adult form. The genes themselves can be copied by template reproduction: one cannot replicate an elephant directly by template reproduction, but one can replicate the instructions for making one and transmit those to the next generation. This is an idea that would have been strange—perhaps incomprehensible—to Darwin, but is less strange to us. We are familiar with the idea that patterns of magnetism on a magnetic tape can carry the instructions for producing a symphony, or that electromagnetic waves can be translated into an image on a television screen. (It is true that the TV image is two-, and not three-dimensional, but that is because of the difficulty of constructing a 3-D screen and not because of any difficulty in transmitting the information.)

The basic picture, then, is that the development of complex organisms depends on the existence of genetic information, which can be copied by template reproduction. Evolution depends on random changes in that genetic information, and the natural selection of those sets of instructions that specify the most successful organisms. But for this to work, the instructions must be interpreted. A compact disc needs a CD player, and the information carried by an electromagnetic wave needs a TV set to convert it into a moving image. Much of modern biology is concerned with how genetic information is translated into form: we describe some of the results below.

What is transmitted from generation to generation is not the adult structure, but a list of instructions for making that structure. As fish evolved into amphibians, or reptiles into birds and mammals, the instructions changed, essentially by random mutation and selection. But the medium in which the instructions were written, and the way in which they were translated into structure, remained basically the same. These processes were wholly unknown to Darwin: they are described below, particularly in Chapters 4 and 10.

But the crucial point we are making in this book is different. As life has become more complicated, the means whereby information is stored and transmitted have also changed: new coding methods have made possible more complex organisms. Thus the coding methods used by fish are not significantly different from those used by birds or mammals: only the message has changed, not the language in which it is written nor the means whereby it is translated. But if we view life on the largest scale, from the first replicating molecules, through simple cells, multicellular organisms, and up to human societies, the means of transmitting information have changed. It is these changes that we have called the 'major transitions': ultimately, they are what made the evolution of complexity possible.

This book is an account of the evolution of complexity. Although essentially Darwinian, it would seem very strange to Darwin, because it is couched in terms of genetic information and how it is stored, transmitted, and translated. This approach to evolution has led us to recognize several 'major transitions', starting with the origin of life and ending with the origin of human language—the most recent change in the way in which information is transmitted between generations. Or perhaps it is not the most recent: we may today be living through yet another major transition, with unpredictable consequences.

What is life?

There are two ways to define life. The first is to say that something is alive if it has certain properties that we associate with living things on Earth; for example, if it grows or responds to stimuli. One problem with this approach is to decide which properties matter. If the first spacemen to land on Mars see an object walking towards them on six legs, with a front end bearing two lenses, a structure resembling a television dish, and a hole surrounded by sharp spikes, they will assume that it is alive, or perhaps an artefact made by creatures that are alive. But if they find that the rocks are covered by a purple slime, they will be less sure. A biologist would probably conclude that the slime was alive if it had a metabolism—that is, if the atoms that compose it were not a permanent part of its structure, but were being taken in from the surrounding environment, combined to form various chemical compounds, and later excreted back to the environment. All things on Earth that are commonly regarded as alive have this property of metabolism, whereas not all of them have legs, eyes, ears, and a mouth. So defining life in terms of metabolism seems sensible. But there are snags. Some things that are not alive have a metabolism of the kind just described, and some things that we might wish to think of as alive do not. We will return to these points later.

An alternative is to define as living any population of entities possessing those properties that are needed if the population is to evolve by natural selection. That is, entities are alive if they have the properties of multiplication, variation, and heredity (or are descended from such entities: a mule cannot multiply, but

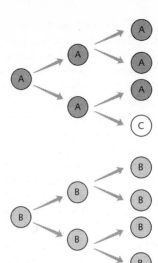

Figure 1.1 The defining characteristics of life. Multiplication implies that one individual can produce two. Heredity requires that there are different kinds of individuals, A, B, and so on, and that individuals produce offspring like themselves. Variation requires that heredity is not perfect, so that occasionally, for example, an A gives rise to a C, as shown. Given these three properties, and an appropriate environment, evolution by natural selection will produce organisms with the other characteristics typical of living things, such as metabolism.

its parents did). The first two are easy to understand. An entity multiplies if it can produce two or more similar entities. Entities vary if, although similar, they are not identical. Heredity is a lot harder (Fig. 1.1). For the present, it is sufficient to say that entities have heredity if there are different kinds—As, Bs, Cs, and so on—and if, when they multiply, As tend to give rise to As, Bs to Bs, and so on: that is, if like begets like. The occasional breakdown of heredity (mutation) gives rise to variation.

Why should we regard these three particular properties as defining life? It is because they are necessary if a population is to evolve all the other characteristics that we associate with life. Multiplication, variation, and heredity are not by themselves sufficient to guarantee the evolution of complex organisms. The environment must also be appropriate. For example, organisms able to walk would not evolve if there was no energy available to them, or if it was too hot to permit the existence of solid structures, or if there was too strong a gravitational field. In other words, evolution of particular structures depends on the environment, and, more generally, on the laws of physics and chemistry. But multiplication, variation, and heredity, even if they do not guarantee evolution, are at least necessary for it.

In this chapter, we explore further the relation between these two concepts of life, the metabolic and the genetic. In the last section we shall come back to a synthesis of these two, essentially complementary, views.

We said earlier that it is possible for an object to have metabolism, but not be alive. The object we had in mind was a fire. Atoms are continuously entering fires, in the fuel or oxygen supply, are involved in a series of chemical changes, and are leaving fires, mainly in carbon dioxide or water molecules. Yet while this is going on a fire may maintain a constant form, like the flame of a Bunsen burner with its blue centre and yellow margin. Fires also multiply. One can use a match to light a Bunsen burner, or even a Bunsen burner to set fire to a laboratory. They also vary in size, shape, and colour. Why, then, do we not think of fires as alive?

One possible answer would be that a fire must be supplied continuously with fuel or it will go out. Clearly this won't do: an animal also will die if it is not fed. One reason for thinking that a fire is not alive is that it is not sufficiently complicated and, in particular, because it does not have organs that ensure its survival and reproduction. Thus our spacemen identified the walking object as alive because it had parts—legs, eyes, ears, mouth—with functions.

We would decide, then, that fire is not alive, but that the walking object on Mars probably is, because fire is too simple and does not have parts ensuring its survival and reproduction. What of the purple slime on the Martian rocks? We have already granted that the smear has metabolism, but so does fire. But if the smear is at all like living things on Earth, it has more than metabolism. Most of the steps occurring in living metabolism depend on enzymes. We will have more to say about enzymes later: for the present it is sufficient to say that they are large complex molecules that help to bring about specific chemical reactions, while themselves remaining unaltered. They are the chemical equivalents of the legs and eyes that persuaded us that the walking object was alive. They are organs that function to bring about the growth of the system as a whole. Specific enzymes could no more come into existence without natural selection than could legs or eyes.

We expect living things to have organs ensuring their growth, survival, and reproduction. We do not think of a fire as being alive because it lacks such organs. This is because fires lack heredity, and so do not evolve by natural selection. Fires vary, but the characteristics of a fire depend only on the supply of fuel and oxygen at the time, and not on whether the fire was lit by a match or a cigarette lighter. Lacking heredity, fire does not evolve, and so lacks the adaptive complexity that only natural selection can confer.

It is true that the object could also be an artefact, designed by intelligent beings. Interestingly, it is not in principle possible to tell the difference between a living organism and a product of intelligent design simply by looking at the object itself. If we are to distinguish between a living being and an artefact, it can only be by knowing its history. We could, if we wished, say that an object (a motorcar or a termite mound) is an artefact if it was constructed by entities (humans, termites) unlike itself, and a living organism if it is the product of evolution, although we would be in difficulty if someone constructed a living cell, indistinguishable from a real cell, in a test-tube.

It follows that the problem of the origin of life is the problem of how entities with multiplication, variation, and heredity could arise, the starting-point being the chemical environment of the primitive Earth. Given these three properties, the other characteristics we expect of living things will evolve. We discuss the origin of life Chapter 3. In this chapter, we make some general points about growth and heredity but, first, we must say a word about the possibility that life originated elsewhere in the Universe, either accidentally or by the deliberate action of intelligent extraterrestrials. This possibility cannot, we think, be ruled out. If it did happen, we are still left with the problem of explaining life's ultimate origin, which is not all that different from explaining its origin on Earth. In the absence of any evidence favouring an extraterrestrial origin, it seems sensible to treat the problem as a terrestrial one.

Autocatalysis

Before there can be heredity there must be reproduction, and before that there must be growth. The essence of growth is autocatalysis (Fig. 1.2). In an auto-catalytic process, a chemical compound A undergoes a series of transformations, $A \rightarrow B \rightarrow C \rightarrow D$, etc. The crucial feature is that the last member of the series, say D, then produces two molecules of A. Thus each A molecule is the starting-point of a new cycle, leading to a new D, and so to two new As. Hence

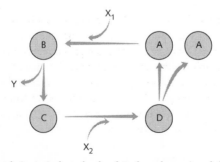

Figure 1.2 Autocatalysis. A single molecule of A, through a series of chemical reactions, produces two molecules of A, X_1 and X_2 are raw materials, and Y is a waste product. Autocatalysis is the basis of growth, but does not constitute a hereditary mechanism, because only one kind of molecule is being multiplied.

the concentrations of all the elements of the cycle increase. One can say that the chemical system is growing. In practice, a cycle of this kind will have side chains: a C will not always produce a D, but something else. Therefore the multiplication factor for each turn of the cycle will be less than two, but provided the flow into the side chains is not too great, the system will grow.

Autocatalytic cycles exist in nature: they are not just abstractions. Moreover, they are found outside living organisms—a point that will become important when we come to discuss the origin of life. As early as 1861, the Russian chemist Alexander Butlerov described a reaction, the formose reaction, in which a solution of formaldehyde and sugars formed additional sugars at an accelerating rate. Both sugars and formaldehyde are simple organic compounds that are readily formed in the 'primitive soup' experiments described in Chapter 3. A second point is that any cycle of the kind shown in Fig. 1.2 requires an input of energy, in this case chemical energy. One cannot have a chemical perpetual motion machine any more than one can have a mechanical one. Again, however, this is not a problem on the primitive Earth. The action of solar radiation on the Earth's atmosphere would readily produce the fuel, including formaldehyde, needed to drive the cycle.

Autocatalytic cycles are important for the origin of life, because they would have produced a rich and varied chemical environment on the primitive Earth. But such cycles do not in themselves provide a mechanism of heredity. For heredity, it is not sufficient that an A should form two As. It is also necessary that, if an A should accidentally change to an A_1, say, then a new cycle, $A_1 \rightarrow B_1 \rightarrow C_1 \rightarrow D_1 \rightarrow 2A_1$, should result. If this did happen, then there would be two states, A and A_1, each capable of reproducing itself: this would indeed be heredity. But in general it will not be so. Instead, A_1 will be the start of a side chain, not producing more A_1, or even more A. The change will lead to degeneration, not to heredity.

It may occasionally be the case, however, that alternative autocatalytic cycles, $A \rightarrow 2A$ and $A_1 \rightarrow 2A_1$, are possible, and that one could form the other by an accidental change, or 'mutation'. Since, almost inevitably, one cycle would be more efficient in utilizing the resources of the environment than the other, one would be 'naturally selected': there would be evolution by natural selection. But it would be a rather boring and limited kind of evolution, from A to A_1, or from A_1 to A. This brings up a crucial distinction between two kinds of heredity, which we call limited and unlimited heredity.

Limited and unlimited heredity

The case we have just described, of two competing autocatalytic cycles, is one of limited heredity. Although there are different kinds of entities, A and A_1, that can replicate, the number of possible kinds is not only finite but rather small. It is an advance over simple multiplication, in which a single type of entity increases in number, but it is not enough for evolution. Continuing evolution requires a system of unlimited heredity, in which an indefinitely large number of structures are each capable of replication. In all existing organisms, heredity depends on 'homologous base pairing' in nucleic acids, usually DNA. Leslie Orgel, a pioneer in the study of the origin of replication, once remarked that, as one traces life back to its origins, features disappear one by one, until one is left with homologous base pairing, like the smile on the face of the Cheshire Cat. The process is so important in the understanding of life that it is described in Fig. 1.3, although it will already be familiar to most readers. Base pairing is a mechanism of unlimited heredity. Given a million base pairs, there are 4 raised to the power one million possible structures that can be replicated, or many times the number of atoms in the Universe. It therefore provides an adequate basis for continued evolution.

A second distinction between kinds of hereditary mechanism is that between modular and holistic heredity. Heredity based on DNA is modular. A DNA molecule consists of many 'modules' (base pairs). If any one of these is changed, and the rest left unaltered, then the descendant molecules are changed just in that one module. In contrast, alternative autocatalytic cycles are an example of limited heredity, and they are holistic. There are no parts that can be changed without changing the whole replicating system. For example, in the case of an autocatalytic cycle, it is not just that A is changed to A_1, but all the intermediates

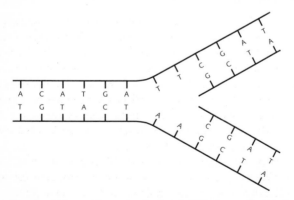

Figure 1.3 Homologous base pairing, and replication. In DNA, there are four kinds of base—adenine (A), cytosine (C), guanine (G), and thymine (T); in RNA, thymine is replaced by uridine. In the two-stranded molecule on the left, A pairs with T, and G with C: it is the specificity of this pairing that is the basis of heredity, and hence of all life. When the molecule replicates, the two strands separate, as shown, and two new molecules, each identical to the original one, are formed by following the pairing rules.

are changed also, B to B_1, C to C_1, and so on. It is for this reason that a holistic system can replicate in relatively few states.

We believe it to be true that all systems of unlimited heredity will turn out to be modular. The statement is true not only of the genetic system based on DNA but also of the only other natural system of unlimited heredity known to us, human language. This is a system in which a small number of unit sounds (phonemes, roughly corresponding to the sounds indicated by the letters of the alphabet) can be strung together in different orders to express an indefinitely large number of different meanings. Changing a single letter changes the meaning of the whole: God's law and Sod's law are not the same, although sometimes one wonders. (We return to language in Chapter 13.) The claim that all systems of unlimited heredity are modular is also true of the many artificial ones, such as the Morse code and the ASCII code.

The mention of human language, or of the ASCII code, raises a query. We started by talking about the replication of structures and ended by referring to systems that transmit information. Much of this book is about the storage and transmission of information. What is the connection between the replication of structures and the transmission of information? Do all structures convey information? Does all information transmission depend on the replication of structures? The notion of information is so central to our purpose that it must be discussed in more detail.

Information and life

In the nineteenth century, society was transformed by machines that convert one form of energy into another. Steam engines convert chemical into mechanical energy, and electric motors and dynamos convert electrical into mechanical energy, and vice versa. These engineering inventions were accompanied by advances in theoretical science, thermodynamics, and electromagnetism. Sometimes the practical applications came first: the steam engine preceded an adequate theory of thermodynamics. Sometimes it was the other way round: Michael Faraday's elucidation of the relation between electricity and magnetism was the foundation of the electrical industry.

Society today is being transformed by machines that convert, not forms of energy, but forms of information. When you speak to a friend on the telephone, information in your mind is converted into a pattern of sound waves in the air, and then into a fluctuating electric current in the telephone wire. If the call is a transatlantic one, these fluctuations are then converted into electromagnetic waves (radio waves), before a series of reverse transformations occurs at the receiver, and your friend acquires the information that started in your mind. Despite the changes in material form from sound waves to fluctuating electric current, and then to radio waves, something is conserved throughout. That something is information. Not only telephones, but tape recorders, record players, radios, television sets, and computers are machines that translate information from one form to another.

These engineering developments have been accompanied by, and have to some extent been influenced by, the mathematical theory of information, although it is probably true that engineering applications have usually led the way. But perhaps the most important impact of information technology on pure science has been in biology, and especially in genetics. The influence is obvious from the terminology of genetics. Code, translation, transcription, message, editing, proof-reading, library, synonymous: these are all technical terms, with quite precise meanings in molecular genetics. The influence of information theory on biology has been in producing a way of looking at things, rather than in providing mathematical tools. One does not need to be a mathematician to work out that, if a code written in four kinds of bases is to specify 20 different amino acids, it must be at least a triplet code (a doublet code could specify only $4 \times 4 = 16$ amino acids). But the way of looking at things is crucial.

Should we think of a gene—that is, a DNA molecule—as a structure that is replicated, or as information that is copied and translated? In present-day organisms, it is both: a gene plays two roles. It acts as a template in gene replication, so that two identical copies are made of a single model. If that were all, the DNA molecule would be just a structure that is replicated. But genes also act to specify the kinds of proteins a cell can make. This process is described in more detail in Chapter 4. For the present, it is sufficient that, via the genetic code, the sequence of bases in a gene determines the sequence of amino acids in a protein (the 'primary structure' of the protein); that this amino acid sequence determines, by the laws of chemistry, the way in which the protein folds up into a three-dimensional structure (the 'tertiary structure' of the protein); and, finally, that this three-dimensional structure determines what the protein does.

By an exact analogy with a tape recorder or a television set, we can say that the sequence of bases in a gene carries information that specifies the structure of a protein, just as the magnetic pattern on the tape specifies the sound that comes out of the recorder. Two points are worth making. First, the process is irreversible. There is nothing odd about this. Some information-transducing machines are reversible. One can use a tape recorder to convert sound into patterns of magnetism on a tape, or vice versa. Other machines are irreversible—you cannot cut a record by shouting at a record player. The process whereby base sequence is converted into protein structure is irreversible. This is the explanation for the non-inheritance of acquired characters, a principle accepted by biologists for 50 years before a chemical explanation for it was available. We will revisit the problem in Chapter 10.

The second point is that translation requires translating machinery. A magnetic tape does not produce sound without a tape recorder, and a gene does not produce a protein without the translating machinery of the cell. We describe this machinery in Chapter 4. But where did it come from? The machinery consists of proteins (and also RNA molecules) whose structure is specified by genes. Unfortunately, this answer leads to an infinite regress: the translating machinery depends on the pre-existence of translating machinery, and so on ad infinitum. Philosophers do not like infinite regresses, but biologists are used to them, in the form of the classic chicken-and-egg problem. We discuss how the translating machinery may have originated in Chapter 4.

We have been describing how things are today. But, as we explain in Chapters 3 and 4, things were not always that way. The first replicating molecules, whether nucleic acids or something simpler, could not have specified anything, and so could not be said to carry information. They are best thought of simply as replicating structures. Only after the evolution of the translating machinery, and hence of specific proteins coded for by genes, is it sensible to talk of genes carrying information. Information theorists use the phrase 'information is data plus meaning'. In biology, the base sequence of nucleic acids provides the data, and the meaning is the structure and function of proteins.

The dual nature of life

We said earlier that there are two ways of defining life: the first is in terms of the complex structure of living organisms and, in particular, their possession of organs adapted to ensure survival and reproduction; the second is by the possession of those properties, especially heredity, needed for evolution by natural selection. We must now try to synthesize these two approaches. Aristotle asserted that life has a dual nature: material is provided by the egg and a formatting force (*entelecheia*) by the sperm. It is unfortunate that he expressed the idea in a manner that could serve as the basis for treating women as second-rate human beings (and, incidentally, in a manner that is in fact false, as far as the genetic contribution of the two sexes is concerned). But in emphasizing the two aspects of life—metabolic and informational—he was correct. As the American geneticist Hermann Muller remarked two millennia later, Aristotle's insight, although nor literally true, would have been worth a Nobel prize.

René Descartes argued that living beings were machines, and could be understood as such. This concept was characteristic of the seventeenth century. It is illustrated by one of the great triumphs of biology: the discovery by William Harvey of the mechanism of blood circulation driven by the heart. All biochemists and molecular biologists today are 'mechanical materialists'. But the machines they study differ from the ones imagined by Descartes. As the philosopher and mathematician Gottfried Leibniz was the first to point out, natural (or 'divine') machines are infinitely divisible. If we analyse a living organism, we find that it is composed of micro-machines (metabolic cycles, enzymes), whereas a steam engine has parts but is not composed of such micro-machines.

Another difference between artefacts and living machines noted by Leibniz is that there is always some *entelecheia* associated with the latter. The nature of this driving force was not further specified, but that it exerted some control on the system was clear to him. Much later, in his famous book *What is life?* (1944), the physicist Erwin Schrödinger emphasized this aspect of life. The book contains the famous remark that the gene, the information-carrying unit of the genetic material, must be an 'aperiodic crystal'. The genetic material must resemble a crystal in being stable and relatively inert, but it must also be 'aperiodic', in the sense of being composed of several different kinds of unit and not just of one kind of unit like a crystal of salt. The reason is that a string of identical units— for example, AAAAA—cannot convey information, whereas a string of dissimilar units can. But Schrödinger also knew that living organisms must *function*. Today, we would express this aspect of life by saying that living systems cannot maintain their active state without a continuous influx of matter and energy.

The contemporary mathematical physicist Freeman Dyson, in a small volume entitled *Origins of life* (1985), revisited the problems posed by Schrödinger. He recognized that life required two things: a self-maintaining metabolic system and genetic material. He thought that concentrating on the latter had not given much insight into the origin of life, and advised that we should concentrate on the former instead. But what does self-maintenance mean? A living system is in continuous change, and some of the changes lead to degradation (that is why biochemists store their compounds in the freezer). It follows that, if the system is to maintain itself, it must be able to overproduce its own material. That is why the metabolic system must be autocatalytic: autocatalysis is needed for self-maintenance, let alone for growth and reproduction.

Some of these ideas were expressed earlier by Tibor Gánti, a Hungarian chemical engineer who became a theoretical biologist. As early as 1966, he argued that life consists of two sub-systems: a homeostatic-metabolic system and the 'main cycle', by which he meant informational control. In *The principle of life*, published in 1971, he described the 'chemoton', the basic design for a minimum chemical system showing all the characteristics of life. Oversimplifying, the chemoton consists of an autocatalytic chemical cycle, and an informational molecule (contained in a bag, so that constituents of the system could not float away in solution). According to this view, viruses are not alive. To use an analogy from computer science, viruses are like programs instructing computers to print them out in as many copies as possible, even if the computer is ruined in the process. It is not the virus, but the cell, that is analogous to the computer. A living being resembles a computer, rather than just a program, although it has its own program as a sub-system.

Gánti also discusses what are the criteria of life. By criteria, he means the empirically determined characteristics essential for life. One might object that some characteristic might be found in all living things on Earth, and yet be accidental. For example, it could have been the case that all organisms on Earth are light blue. How would you know whether blueness is necessary to life, or accidental? The empirical approach to the definition of life, therefore, runs the risk that some accidental trait will be accepted as essential. This should not worry us too much. All natural sciences have an empirical basis, and are therefore subject to modification when new data are discovered.

Gánti, then, adopted an empirical approach to identifying the defining criteria of life. He distinguished two types of criteria: 'absolute' and 'potential'. By an absolute criterion, he meant one necessarily present in all living creatures. A potential criterion is one that is not necessarily present in all living things, but that is necessary if organisms are to reproduce and evolve (the term 'potentiating', rather than 'potential', might express the distinction better). For example, mules are alive, but cannot reproduce. The ability to reproduce, therefore, is a potential but not an absolute criterion of life. Without going into the details of Gánti's argument, it is impressive that, very early, he recognized that both metabolism and informational control are necessary. Because we are interested in evolution, and not merely in the survival of individuals, we will concentrate on the potential criteria—in particular, on multiplication and heredity.

CHAPTER 2

THE MAJOR TRANSITIONS

The theory of evolution by natural selection does not predict that organisms will get more complex. It predicts only that they will get better at surviving and reproducing in the current environment, or at least that they will not get worse. Empirically, many and perhaps most lineages change little for many millions of years. As D. M. S. Watson, who taught one of us zoology, once remarked, crocodiles have done damn all since the Cretaceous. Lampshells and horsetails have done damn all for longer than that.

Yet some lineages have become more complex. There is some sense in which elephants are more complex than slime moulds, and oak trees more complex than green algae, even if we find it hard to say just what that sense is. One approach to definition is in terms of the number of parts composing an organism, or the number of behaviours possible to it. An elephant has many different kinds of cells—at least several hundred—whereas a slime mould has very few. It is also capable of a large number of behaviours—walking, wallowing, nursing its young, uprooting trees, trumpetting, and so on—whereas a slime mould, again, has few. This is true, but not very helpful, partly because it is hard to quantify, and partly because it does not readily lead to additional questions.

A more fruitful approach to measuring complexity comes from mathematics. The American mathematician G. J. Chaitin has suggested that we can measure the complexity of a structure by the length of the shortest list of instructions that will generate it: the complexity of a cake is measured by the length of the recipe telling you how to make it. Unfortunately, although we can say precisely how long a list of instructions (that is, bases in the DNA) is needed to generate a protein, or set of proteins, we have no idea of the minimum number of base pairs needed to make an elephant. What we can say, however, is approximately how many base pairs are actually used. Even this simpler question is not quite straightforward. We cannot just measure the DNA content of the nucleus of the fertilized egg of an elephant (which we should halve, because the egg contains two almost identical copies of the same information, one from each parent, and two copies of a message do not carry more information than one). The snag is that much of the DNA of any higher organism does not contribute useful information: it is like the static in a poor radio message. We discuss such DNA briefly

Table 2.1 Numbers of genes in various organisms

Species	Type	Approximate gene number
Prokaryotes		
Escherichia coli	Bacterium	4000
Eukaryotes (except vertebrates)		
Oxytrochis similis	Ciliated protozoan	12 000–15 000
Saccharomyces cerevisiae	Yeast	7000
Dictoyostelium discoideum	Slime mould	12 500
Caenorhabditis elegans	Nematode	17 800
Drosophila melanogaster	Insect	12 000–16 000
Strongylocentrotus purpuratus	Echinoderm	<25 000
Vertebrates		
Fugu rubripes	Fish	50 000–100 000
Mus musculus	Mammal	80 000
Homo sapiens	Mammal	60 000–80 000

in Chapter 8. But if it is not needed it should certainly not be included in any measure of complexity.

A rough idea of the amount of informative DNA in organisms varying in apparent complexity is given in Table 2.1. The transition between prokaryotes (essentially, bacteria) and eukaryotes (all the rest) is discussed in Chapter 6, where we also discuss the reasons for substantial increase in informative DNA. It is less clear why vertebrates appear to have more informative DNA than invertebrates. As vertebrates ourselves, we are perhaps less surprised by the difference than we should be: why should an insect, with legs and wings, need less DNA than a fish?

It seems, then, that although there is no general reason why evolution should lead to greater complexity, it has in fact done so in some cases. In the next section, we argue that this increase has depended on a small number of major changes in the way in which information is stored, transmitted, and translated. These changes we refer to as the major transitions.

The major transitions

The easiest way of explaining what we mean by the major transitions is to list them (Table 2.2). The brief explanation of this list that now follows is in effect a synopsis of the rest of the book: if some statements seem obscure, we hope that they will be made clearer in the appropriate chapter.

1. *Replicating molecules→populations of molecules in compartments*. We think that the first objects with the properties of multiplication, variation, and heredity

Table 2.2 The major transitions

Replicating molecules	Populations of molecules in protocells
Independent replicators	Chromosomes
RNA as gene and enzyme	DNA genes, protein enzymes
Bacterial cells (prokaryotes)	Cells with nuclei and organelles (eukaryotes)
Asexual clones	Sexual populations
Single-celled organisms	Animals, plants, and fungi
Solitary individuals	Colonies with non-reproductive castes (ants, bees, and termites)
Primate societies	Human societies (language)

were replicating molecules, similar to RNA but perhaps simpler, capable of replication, but not informational because they did not specify other structures. If evolution was to proceed further, it was necessary that different kinds of replicating molecule should co-operate, each producing effects helping the replication of others. We argue that, if this was to happen, populations of molecules had to be enclosed within some kind of membrane, or 'compartment'.

2. *Independent replicators→chromosomes*. In existing organisms, replicating molecules, or genes, are linked together end to end to form chromosomes (a single chromosome per cell in most simple organisms). This has the effect that when one gene is replicated, all are. This co-ordinated replication prevents competition between genes within a compartment, and forces co-operation on them. They are all in the same boat. We discuss this transition in Chapter 5.

3. *RNA as gene and enzyme→DNA and protein*. There is today a division of labour between two classes of molecule: nucleic acids (DNA and RNA) that store and transmit information, and proteins that catalyse chemical reactions and form much of the structure of the body (for example, muscle, tendon, hair). It seems increasingly plausible that there was at first no such division of labour and that RNA molecules performed both functions. The transition from an 'RNA world' to a world of DNA and protein required the evolution of the genetic code, whereby base sequence determines protein structure. This is the topic of Chapter 4.

4. *Prokaryote→eukaryote*. Cells can be divided into two main kinds. Prokaryotes lack a nucleus, and have (usually) a single circular chromosome. They include the bacteria and cyanobacteria (blue-green algae). Eukaryotes have a nucleus containing rod-shaped chromosomes and usually other intracellular structures called 'organelles', including the mitochondria and chloroplasts described on pp. 70–77. The eukaryotes include all other cellular organisms, from the single-celled *Amoeba* and *Chlamydomonas* up to humans. We discuss the transition from prokaryotes to eukaryotes in Chapter 6.

5. *Asexual clones→sexual populations.* In prokaryotes, and in some eukaryotes, new individuals arise only by the division of a single cell into two. In most eukaryotes, in contrast, this process of multiplication by cell division is occasionally interrupted by a process in which a new individual arises by the fusion of two sex cells, or gametes, produced by different individuals. Although familiar, this transition is one of the most puzzling; we discuss it in Chapter 7.

6. *Protists→animals, plants, and fungi.* Animals are composed of many different kinds of cells—muscle cells, nerve cells, epithelial cells, and so on. The same is true of plants and fungi. Each individual, therefore, carries not one copy of the genetic information (two in a diploid) but many millions of copies. The problem, of course, is that although all the cells contain the same information, they are very different in shape, composition, and function. In contrast, protists exist either as single cells, or as colonies of cells of only one or a very few kinds. How do cells with the same information become different? How do different kinds of cells come to be arranged so as to form the adult structure? What problems had to be solved before animals and plants could evolve? We discuss these questions in Chapter 10.

7. *Solitary individuals→colonies.* Some animals, notably ants, bees, wasps, and termites, live in colonies in which only a few individuals reproduce. Such a colony has been likened to a superorganism, analogous to a multicellular organism. The sterile workers are analogous to the body cells of an individual, and the reproducing individuals to the cells of the germ line. The origin of such colonies is important; it has been estimated that one-third of the animal biomass of the Amazon rain forest consists of ants and termites, and much the same is probably true of other habitats. It is also interesting for the light it sheds on the origin of human societies. We discuss these origins in Chapter 11.

8. *Primate societies→human societies, and the origin of language.* We argue in Chapter 12 that the decisive step in the transition from ape to human society was the origin of language. We have already emphasized the similarities between human language and the genetic code. They are the two natural systems providing unlimited heredity. The nature and origin of human societies are the topic of Chapter 12, and in Chapter 13 we discuss the origin of language.

Because we are concerned with information, we should perhaps have included in our list the evolution of a nervous system capable of acquiring information about the external world, and using that information to modify behaviour. Certainly the acquisition of a nervous system was a necessary precondition for the subsequent evolution of language. Our only excuse for omitting it is one of incompetence!

Of the eight transitions that we have listed, we think that all but two were unique, occurring just once in a single lineage. The two exceptions are the origins of multicellular organisms, which happened three times, and of colonial animals with sterile castes, which has happened many times. There are interesting implications of the occurrence of six unique transitions, together with the origin of life itself, which we also think to have been a unique sequence of events. Any one of them might not have happened, and if not, we would not be here, nor any organism remotely like us.

A common problem

One reason for discussing events as different as the origin of the genetic code, of sex, and of language in a single book is that we think that there are similarities between the different transitions, so that understanding one of them may shed light on the others. One feature in particular crops up repeatedly. Entities that were capable of independent replication before the transition could afterwards replicate only as part of a larger whole. For example:

1. It is now generally accepted that, in the origin of the eukaryotes, an important event was the symbiotic union of two or more different kinds of prokaryotes, which could once replicate independently, but can now replicate only when the whole cell replicates.

2. After the origin of sex, individuals can reproduce only as members of a sexual population, whereas earlier they could reproduce asexually, on their own.

3. The cells of a higher organism, plant or animal, can divide during growth, but their long-term future (or rather, the long-term future of their genes) depends on being part of a multicellular organism.

4. Ants, even the reproductive castes, can reproduce only as members of a large colony, but their ancestors could reproduce as members of a sexual pair. It is effectively true, also, that humans can reproduce only as part of a larger social group.

This common feature of many of the transitions raises a common problem. Why did not selection between entities at the lower level (in the examples above, between prokaryotic cells, asexual individuals, individual protist cells, individual ants) not disrupt integration at the higher level (eukaryotic cell, sexual population, multicellular organism, ant colony)? In trying to answer this question, it is not sufficient to point to advantages possessed by the higher-level entity. An ant colony may be very efficient at exploiting the environment, but that does not explain why an individual ant should sacrifice its chances of reproduction to help the colony. An adequate account requires that we explain the origin of the higher-level entity in terms of selection acting on the lower-level entities.

This problem is not a new one: it is often referred to as the 'levels of selection' problem. It has been around since Darwin, but the modern debate was initiated by the publication, in 1962, of *Animal dispersion*, a book by the ornithologist V. C. Wynne-Edwards. He asked the following question: given the enormous reproductive capacity of most animals, why do not their numbers increase until the food runs out, and they starve? His answer was that the increase in numbers is usually limited by their behaviour, before starvation intervenes. As an ornithologist, he pointed to the fact that individuals that do not establish a territory refrain from breeding (this was not the only example in his book, which is full of fascinating natural history). He went on to point out that such behaviour, although beneficial to the population by preventing starvation, is harmful to the individuals that do not breed. Therefore, he argued, such self-sacrificing behaviour requires 'group selection': it evolves because populations whose members display such behaviour survive, whereas populations composed of selfish individuals die out. In other words, such behaviour evolves because selection favours some populations at the expense of others, and not some individuals.

The great merit of Wynne-Edward's book was that it made explicit the idea that, if one claims that some characteristic exists because it is good for the group, or species, then one is assuming that selection acts between species, and not between individuals. The difficulty is that if individuals with some trait are favoured by selection, but groups of individuals with that trait are selected against, then in most cases selection at the individual level will win: if, when food begins to run short, some individuals go on breeding, whereas others refrain, it is those that breed that will transmit their genes to future generations, even if in the long run the result is the extinction of the population. In fact, most of the examples given by Wynne-Edwards could be explained by selection acting on individuals. For example, individuals that do not establish a territory, and do not breed, are not sacrificing themselves for the good of the population: they are making the best of a bad job. There is no need to invoke group selection.

The 'levels of selection' argument first came into prominence as a debate about whether selection acts between individual organisms, or between populations of organisms. In this book we are concerned with a number of levels—genes, chromosomes, cells, organisms, sexual populations, and societies. For us, therefore, the conflict between different levels of selection is crucial. We have to explain how complex entities evolved, despite selection between their components favouring selfish behaviour.

Once a compound entity has been in existence for a long time, in evolutionary terms, it may no longer be possible for its components to return to their ancestral state of independence. A cancer cell may gain a short-term selective advantage over its better-behaved neighbours, and multiply accordingly, but it has no long-term future as a free-living protist. The difficulty of explaining the origin of compound entities, each with its own genetic information, can be illustrated, however, by pointing to conflicts that still exist today between different components of a single organism. Consider the following four examples:

1. Because genes are parts of chromosomes, and chromosomes obey Mendel's laws, sharing equal chances of transmission to gametes and hence to the next generation, one might guess that there was nothing that a gene could do to increase its own chance of transmission relative to other genes, at the expense of organism fitness. Such a guess would be quite wrong. In Chapter 8 we describe various ways in which genes can cheat, gaining excess representation in future generations.

2. It is conventional wisdom that worker bees sacrifice themselves for the good of the colony. This is not always true: various ways in which they cheat are described on p. 127.

3. On pp. 102–6, we describe various ways in which intracellular organelles gain increased representation in future generations, sometimes at the expense of the fitness of the organism.

4. In both Britain and Hungary, some people illegally avoid paying taxes.

If such conflicts remain even after a long period of coexistence, they must have been still more apparent during the evolutionary origins of compound entities. We next consider three reasons why the major transitions, although difficult, were not impossible.

Some possible solutions

Genetic similarity

It is a familiar but curious fact that, almost always, complex multicellular organisms originate from a single cell. This requires that cell differentiation be achieved over again in each generation. It is not the way an engineer would do it. Instead, one would make a little homunculus, putting together groups of already differentiated cells in the appropriate way. But the familiar method has one very important consequence. It ensures that the genes in all the cells of an individual are identical, except for somatic mutations that occurred since the fertilized egg was formed.

The effect of this identity of the genes within an organism is to make it more likely that natural selection will favour genes that cause co-operative rather than selfish behaviour. Thus imagine two alternative genes that could be in a kidney cell. As a shorthand, we will call them co-operative and selfish. The normal, co-operative gene causes the kidney cell to perform its usual function in excretion, whereas the selfish gene causes the cell to de-differentiate, and to enter the blood stream and travel to the gonad, where it has a chance of being transmitted to the next generation—at the expense of reducing the efficiency of the kidney, and hence the survival or fertility of the individual. Which gene will increase in frequency? When we speak of 'a gene increasing in frequency', what we mean is that there will be more copies of genes carrying the same information. It is the information that matters, not the physical object. More copies of a co-operative gene will be passed on, because an organism with a co-operative gene has more offspring, all of which receive copies of the co-operative gene (or half of them, if we allow for the presence of two sets of genes in each cell).

This is essentially the argument proposed by Oxford zoologist William D. Hamilton to explain the evolution of social behaviour. He dealt with a harder problem, the evolution of co-operation between individuals that are not genetically identical, but genetically related. His conclusions can be summarized by the famous inequality, stating that co-operation will spread if $rb > c$, where b is the benefit conferred, c the cost to the benefactor, and r a measure of relatedness. For the cells of a single individual, $r = 1$.

The first facilitating condition for the evolution of co-operation between initially independent replicators, then, is relatedness. This is brought about if a group of interacting individuals are all recently descended from a single ancestor (and so have the same genetic information) or from a small group of ancestors. We have already mentioned two examples that are relevant to the major transitions. Transition 6, the origin of multicellular organisms, probably required that each new individual developed from a single cell. The origin of animal societies required that a new colony be founded by a few individuals: today, most insect colonies are founded by single females. We will mention one other. During transition 4, the origin of the eukaryotes, cells came to contain new organelles, the mitochondria, descended from once free-living bacteria, and still containing some of the original bacterial genes. Competition between the mitochondria in a single cell, and the consequent evolution of selfish mitochondria, is largely suppressed because all the mitochondria in an individual are genetically identical. This is brought about because, in sexual reproduction, mitochondria are inherited from only one parent—in animals, from the mother —and there are only few in the egg. This reduces the danger of selfish mitochondria evolving, but, as we will see, it does not wholly prevent it.

Synergy

Co-operation will not evolve unless it pays. Two co-operating individuals must do better than they would if each acted on its own. One reason why insects become social is that, in a group of co-operatively nesting females, some can forage while others protect the nests against parasitoids, which are a major cause of mortality in the larvae. Co-operative breeding between a mated pair of birds has a similar explanation: one can incubate the eggs while the other forages. In effect, the principle is that of the division of labour. Behavioural examples are easy to think of, but the principle is relevant at all levels. In the RNA world, the same kind of molecules acted as enzymes and as carriers of heredity: transition 3 occurred because it is more efficient to separate these functions between proteins and DNA, respectively. In multicellular organisms, cells can be specialized to perform different functions.

The American computer scientist Peter Corning, in a book called *The synergism hypothesis* published in 1983, reviewed the role of synergy in social and biological evolution. We had not seen his book when we wrote *The major transitions in evolution*, but are happy to acknowledge that he foreshadowed this part of our argument, often using the same examples.

Central control

In human societies, co-operation is often enforced by some form of central authority. Most Englishmen and Hungarians pay taxes because they are punished if they do not. Is there any analogous process in biological systems?

There are two possible forms such central control can take. The first is illustrated in Fig. 2.1. Most higher plants are hermaphrodites, producing both seeds and pollen. As in animals, the mitochondria are transmitted only in the egg cell. Taking a gene's eye view, what would you do if you were a gene in a mitochondrion? As a gene, what you would want to do (that is, what the gene would be selected to do) is to cause more copies of yourself to be present in future generations. If it was possible, you might cause the abortion of the male organs of the flower, because a plant that does not use resources in making pollen can produce more seeds, and so would produce more copies of you. As it happens, several cases are known in which male sterility in plants is caused by genes in the mitochondria: a case that has been extensively studied is in *Thymus vulgaris*, the wild thyme. Now look at the situation from the viewpoint of a chromosomal gene, which can be transmitted in seed or pollen. Male sterility is not in the interest of such a gene. It is therefore understandable that there are chromosomal genes in *Thymus* that suppress the effects of the mitochondrial genes, and restore male fertility.

One can see this simply as an example of conflict between genes, but one can also see it as an example of central control. There are many more chromosomal than mitochondrial genes, so it is perhaps not surprising that for each mitochondrial gene that can mutate to cause sterility, there is at least one chromosomal gene that can suppress it. The American biologist Egbert Leigh called this

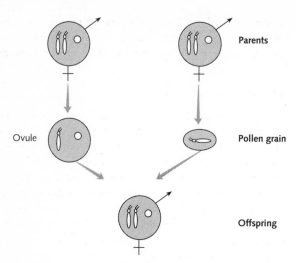

Figure 2.1 Reproduction in a hermaphrodite plant. The offspring receives one copy of each kind of chromosome from each parent (only one kind is shown), but receives its mitochondria, including a circular chromosome, only from the ovule parent. Since a gene in a mitochondrion is transmitted only in ovules, it would pay it to suppress pollen production, so that there are more resources available for ovules. In some plants, for example *Thymus*, genes on the mitochondrion do in fact suppress pollen production, but their action is often suppressed by genes on the nuclear chromosomes. This can be seen as a kind of 'central control', or 'policing', by nuclear genes.

effect the 'parliament of the genes'. Parliaments are ruled by majority voting. Leigh's idea is that, for most possible kinds of selfish gene, there will be many more genes whose interests would be served by suppressing the selfish behaviour. Of course, no vote is taken: it is just a matter of the number of possible kinds of mutation. We return to this topic in Chapter 8.

Another, rather different phenomenon can also be seen as central control. Again, we will explain it by an example. Lichens are a symbiotic union of a fungus and an alga. The algal species involved can live independently but may be engulfed by the fungus to form a lichen association. Should we think of this as an example of slavery or of co-operation? How can we decide? If co-operation is the appropriate image, then both partners should benefit, and we would therefore expect them to have characteristics that facilitate the symbiotic union. In most lichen associations, nobody has been able to point to characteristics of the algae that look as if they evolved to encourage association with the fungus. It is hard to be sure, but it may be that slavery rather than co-operation is the appropriate image; symbiotic associations need to be looked at carefully with this question in mind.

In later chapters we look in detail at the various transitions in which independent entities have come to coexist. Usually, both relatedness and synergy were important. Occasionally, central control may also have been relevant.

No foresight, and no way back

There are two other features of the transitions that need emphasizing. The first is that evolution by natural selection lacks foresight. A transition may have opened up new possibilities for future evolution, but that is not why it happened. For example, the origin of eukaryotes from prokaryotes involved major changes in chromosome structure, and in the way in which one copy of each chromosome is passed to each daughter cell when the cell divides. We describe the changes in some detail in Chapter 6 . We argue that, before the changes, there was a serious constraint on the total amount of DNA that could be replicated, which limits the maximum DNA content of prokaryotes. After the changes, this constraint was lifted, permitting a further increase in complexity. But the changes did not occur *because* they removed the constraint. If we are right, the changes were forced on the early eukaryotes because of the loss of the rigid outer cell wall of prokaryotes. This pattern, of a change occurring for one reason but having profound effects for other reasons, is often repeated.

The other feature is the difficulty of reversing the transitions, once they had happened. This can be nicely illustrated by the example of sex. It is a surprising fact that no gymnosperm (coniferous tree) has ever reverted from sexual reproduction to parthenogenesis. The explanation is simple. We mentioned earlier that, in sexual reproduction, intracellular organelles are usually transmitted by one parent only. In gymnosperms, the chloroplasts (organelles that carry out photosynthesis) are transmitted only in the pollen. Parthenogenetically produced seeds, therefore, would give rise to colourless seedlings that could not grow.

It turns out that there are many similar obstacles that must be overcome before a sexual organism can revert to parthenogenesis. Once sex had arisen, many secondary adaptations became associated with it, so that sex is hard to abandon. As it happens, no mammalian egg will develop without fertilization, either in the wild or in the laboratory. The reason is understood, but it is complicated: it certainly has nothing to do with the reasons why sex evolved in the first place. Sometimes, however, sexual populations can have parthenogenetic descendants. Indeed, there are many parthenogens in nature, including many flowering plants, and animals as complicated as lizards. In the case of most other transitions, however, irreversibility seems absolute. Multicellular organisms never have single-celled descendants; eukaryotes never have prokaryotic descendants; it is unclear whether colonial insects, with sterile castes, have ever had solitary descendants.

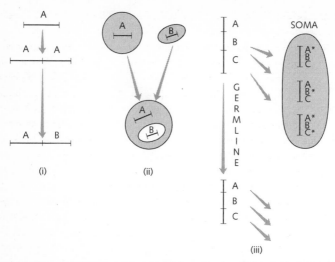

Figure 2.2 Three ways of increasing the genetic information in a single individual. (i) Duplication and divergence: a single gene duplicates, and then the two copies diverge in sequence. (ii) Symbiosis: two individuals, with different ancestry and genetic information, fuse (usually by one being swallowed by the other). (iii) Epigenesis: in multicellular organisms, the complete genetic message, ABC, etc., is transmitted in the 'germ line', from fertilized egg to the gametes, which fuse to form the next generation. During development, the complete message is transmitted to all somatic cells, but different parts of the message are active in different cells (* indicates genes that are active).

How did genetic information increase?

Despite our difficulty in saying just how much DNA is needed to specify an elephant, it is certainly more than that needed to specify a bacterium. Figure 2.2 shows three ways in which the information in a lineage can increase: by duplication and divergence, symbiosis, and epigenesis.

Duplication and divergence

The simplest process is the duplication of a piece of DNA, which can vary in length from a single gene to a whole set of chromosomes. Such accidental events are not all that infrequent. In itself, a duplication does not add to the total quantity of information present: two copies of a message are not more informative than one. All it does is to produce additional DNA that can later be programmed by selection. It is worth noting, however, that the procedure is rather different from the way in which one might add memory to a computer. In the latter case, the additional memory would initially be blank (unless one added an already programmed chip). In evolution, the new DNA already carries a message, albeit a redundant one. New information requires that this message be altered step by step.

We know that the duplication of genes has been important. A classic example concerns haemoglobin, the protein that carries oxygen in the blood. It is a compound of four subunits, of two kinds, each kind programmed by a different gene. The two genes arose by duplication, followed by minor divergence. A further round of duplication and divergence produced the different haemoglobin in the fetus of mammals. Gene duplication is common, but does not always lead to an increase in information: more often, one of the two copies degenerates, because natural selection does not maintain two copies if one will do. Our chromosomes are full of such fossil genes, so-called pseudogenes. It is only occasionally that the duplicate copy acquires a new function.

The important point is that duplication, whether of single genes or whole genomes, does not in itself produce significant novelty. It merely provides additional DNA that is not needed, and so can be programmed to perform new functions. It does not cause increased complexity, but it does provide the raw material for such an increase to occur later.

Symbiosis

Symbiosis is the process whereby two different kinds of individual come to live together. Most symbioses are parasitic in nature; one individual benefiting at the expense of the other. Here, however, we are interested in mutualistic symbiosis, in which both partners benefit. In Chapter 9, we discuss the relevance of symbiosis for the origin of eukaryotic cells. In Fig. 2.3, we illustrate the role of symbiosis in two earlier transitions, in which different independently

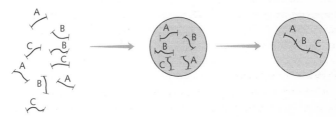

Figure 2.3 Symbiosis in the early evolution of life. Initially, different replicating molecules, A, B, and C, were free in solution, or perhaps bound to a surface. They would have competed for 'resources'—that is, for the small molecules of which they were composed. Later, different replicators would have been contained within a membrane, or protocell. If the growth and division of the protocell depended on the number of molecules it contained, then there would be selection for some degree of co-operation. Finally, different molecules were joined end to end to form a 'chromosome'. Then, one molecule could replicate only if all did, and co-operation would be more strongly selected.

replicating molecules were first enclosed within a cell membrane, and then linked end to end to form a chromosome. Symbiosis differs from duplication in that there is an immediate increase in the genetic information within an individual. The new individual has the sum total of the information present in the two symbionts, although some of the information may prove to be redundant, and later be lost.

Epigenesis

Fibroblasts, liver cells, and epithelial cells are different; and the differences are inherited. If one establishes a tissue culture of fibroblasts, the cells will divide many times, but their descendants are still fibroblasts. In the same way, the descendants of epithelial cells are still epithelial cells, and so on. How does such heredity work?

The first biologist clearly to understand that inheritance involves information was August Weismann. He rejected the then generally accepted idea that acquired characters are inherited because he could not see how, for example, the blacksmith's muscles could so influence his sperm that his sons would also develop big muscles. He wrote, in *The evolution theory* (1902; we quote the English translation of 1904), that to suppose this is 'very like supposing that an English telegram to China is there received in the Chinese language'. He was therefore puzzled about how the cells of the body could be so different from one another. There were, he thought, two possible explanations. The idea that he favoured was that the fertilized egg contains a complete set of genes (he called them ids), but that during development the kidney cells received only those ids needed in the kidney, epithelial cells only those ids needed in the epithelium, and so on. Only the germ line retained a complete set of ids. He did see, however, that an alternative explanation is possible. Each kind of cell receives a complete set of genes, but becomes different because it receives different external stimuli, activating different genes. This idea, which we today accept as correct, he rejected because of the large number of specific stimuli that it demands. It was an excusable mistake. Today, a hundred years later, we are only just beginning to understand the nature of these stimuli.

What happens in most multicellular organisms is this. With a few exceptions, every cell receives a complete set of genes but different genes are active in different cells. This state of activation is transmitted to daughter cells when a cell divides. This is a new kind of inheritance, epigenetic inheritance, that does not depend on differences in the base sequence of DNA. The mechanism is described in more detail in Chapter 9.

An embryo, then, has a dual inheritance system, one system depending on the copying of DNA base sequences, the other on the copying of states of gene activity. There is an obvious analogy between the differentiated cells of an animal body, the various castes in an ant colony, and the different trades and professions in human society. The Israeli biologist, Eva Jablonka, has pointed out that the analogy between an animal body and human society is deeper than just the presence of differentiated parts. Human society also depends on a dual inheritance system, based on DNA and on language.

CHAPTER 3

FROM CHEMISTRY TO HEREDITY

The last sentence of Darwin's *Origin of species* reads 'There is a grandeur in this view of life, with its several powers, being originally breathed by the Creator into a few forms or into one; and that, while this planet has gone cycling on according to the fixed law of gravity, from so simple a beginning endless forms most beautiful and most wonderful have been, and are being evolved'. Curiously, the phrase 'by the Creator' crept in only in the second edition: it was absent in the first. It seems that Darwin, perhaps to please his wife Emma, was willing to leave the problem of the origin of life, a problem he had little prospect of making progress with, as a matter for religious interpretation. In contrast, the problem of the origin of man was one he believed he could usefully think about.

The primitive soup

The first serious scientific attacks on the origin of life came from the Russian biochemist A. I. Oparin in 1924 and from J. B. S. Haldane in 1929. They argued that, if the primitive atmosphere lacked free oxygen, a wide range of organic compounds might be synthesized, using energy from ultraviolet light and lightning discharges. The absence of oxygen was important, because otherwise any organic compounds formed would rapidly be oxidized to carbon dioxide (CO_2) and water (H_2O). Haldane suggested that, in the absence of living organisms to feed on the organic compounds, the sea could have reached the consistency of a hot dilute soup.

In 1953, Stanley Miller, on the advice of Harold Urey, tested the idea by passing an electric discharge through a chamber containing water, methane (CH_4), and ammonia (NH_3). The results were dramatic. In these and similar experiments, a wide range of organic compounds have been produced, including many of the amino acids of which proteins are made, a variety of sugars, and various purines and pyrimidines (components of the nucleotides of which RNA and DNA are made).

Although these experiments were enormously encouraging, there were snags. Some essential molecules were present in low concentrations, or altogether absent. The sugar ribose, which forms the backbone of RNA and DNA, is produced, but the yield is low. Long-chain fatty acids, needed to form biological membranes, were absent. Perhaps more fundamental is the question of how these simple organic molecules could be strung together to form biological polymers. For example, proteins consist of strings of amino acids joined by a particular chemical bond, the peptide bond. Non-chemists can think of a protein as a poppet necklace, with different kinds of beads joined together in a precise order by a particular kind of junction. Sidney Fox found that by heating and drying a mixture of amino acids, and then dissolving the mixture in water, he could obtain strings of amino acids displaying weak catalytic activity. Unfortunately, however, the amino acids were linked together in a variety of different ways, and not only by peptide bonds. Similar difficulties arise for a second class of polymer, the nucleic acids. RNA and DNA are polymers formed of nucleotides. Even if nucleotides could be synthesized in Miller-type experiments, and this is not easy, it is not clear how they could be linked in the appropriate way. If some of the linkages were of the wrong kind, replication would be blocked.

To summarize, there is no difficulty in seeing how a wide range of organic compounds could have been formed abiotically, including many of the compounds important in modern organisms. But there is a lack of specificity in the reactions that take place, and a particular difficulty in understanding how polymers (proteins, nucleic acids), linked by specific chemical bonds, could have been formed.

The primitive pizza

A possible way out of these difficulties has recently been suggested by Günter Wächtershäuser, a somewhat unusual contributor to the field. Although he has a doctorate in chemistry, he is not working in an institute or university. He is a patent lawyer for chemical patents in Munich. (It is amusing that Einstein was also working at a patent office when he developed the special theory of relativity.) Before Wächtershäuser's first publication on the origin of life in the late 1980s, he had been strongly influenced by the philosophy of Karl Popper. This has led him to take the idea of testable hypotheses seriously, and to be disgusted by the, sadly rather common, theoretical sloppiness in this rather unconventional field.

The idea is that reactions may have taken place between ions bonded to a charged surface (for non-chemists, an ion is an electrically charged atom or molecule). For example, in the solid state common salt is not electrically charged, but in solution in water the molecules break up into positively charged sodium ions and negatively charged chlorine ions. Many important organic compounds ionize in solution: an example is the phosphate ion, PO_4^{3-}, which is important both in nucleic acids and in energy metabolism. Because unlike charges attract one another, ions in solution will become attached to charged surfaces. They are then free to move slowly across the surface, while maintaining a constant orientation. This would greatly increase both the speed and the specificity of the chemical reactions occurring. Because of the importance of negatively charged ions, such as the phosphate ion, in biochemistry, Wächtershäuser suggests that the relevant surface was the positively charged iron pyrites, or fool's gold.

Wächtershäuser has suggested some very specific chemical reactions that could have taken place; his ideas now need to be tested experimentally. One reason why surface bonding is important is that the molecules are held in a particular orientation, and are free to move only in a single dimension. If some of them were upside down, or all of them were free to move in three dimensions, they would never link together. Binding to a surface would also increase the local concentration of interacting molecules, and so speed up the reaction. Equally important, it would ensure that reacting molecules were held in a given orientation relative to one another, and so increase the specificity of the reactions.

The origin of replication

The first artificial replicator, not needing enzymes for its replication, was synthesized by K. von Kiedrowski in 1986. The object that was replicated was a small piece of DNA, of six base pairs: the 'units' from which new copies were made were two single-stranded molecules, each consisting of three bases linked end to end. This was an important achievement, but not yet a solution to our problem. It is a system with limited heredity, whereas we need a system with unlimited heredity. In all probability, this means a system replicating by homologous base pairing. As yet nobody has devised such a system, composed of monomers that could have arisen in the absence of living organisms and able to replicate without enzymes.

The simplest existing replicating system is shown in Fig. 3.1. This is a real example of evolution in a test-tube, with no cells present. RNA molecules appear that are better and better at getting themselves replicated in this environment. It is a beautiful demonstration of the power of natural selection to generate adaptations that would never appear by chance alone. Given the right circumstances, the end point of evolution is a unique RNA molecule 235 bases long, regardless of the starting sequence. But despite this demonstration of the power of natural selection to generate unlikely adaptations, the example is in a sense cheating, if we want to explain the origin of the first replicators. The system works only because the right monomers, and more important, a complex enzyme, the Q_β replicase, are supplied. No such enzyme could have existed on the primitive Earth, before the origin of life.

One difficulty in understanding the non-enzymatic replication of a nucleic

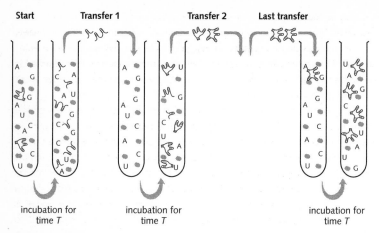

Figure 3.1 Evolution in a test-tube. The initial tube contains the four nucleotides from which RNA is synthesized, a replicase enzyme that copies RNA, and a 'primer' RNA molecule. This primer molecule is copied many times, with some errors. After incubation for time T, a drop of solution is transferred to a new tube containing nucleotides and enzyme (but not, of course, any primer). The process of incubation and transfer can be repeated indefinitely, and evolutionary changes in the population of RNA molecules observed.

acid-like molecule lies in explaining the specificity with which the units, the monomers, are linked together. If they are linked by the wrong chemical bonds, replication will be blocked. At present, the best hope seems to lie in seeking a polymer with a chemically simpler backbone than RNA, thus reducing the number of ways in which the monomers can be linked. The extra specificity to be gained from metabolism on a surface may also help. For the present, we have to accept that the origin of molecules with unlimited heredity is an unsolved problem. But progress is rapid. Despite the last sentence of the *Origin of species*, we cannot advise creationists to put their faith in the belief that only God could create a molecule with unlimited heredity.

The accuracy of replication and the error threshold

Replication is not perfect. If it were, there would be no variation for selection to act on. But initially the problem would have been too much mutation, and not too little. Most mutations reduce fitness. Selection is therefore needed to maintain a meaningful message. The old game of Chinese whispers demonstrates that, without selection, the result is chaos. How accurate must replication be? Imagine a message—for example, a DNA molecule—that replicates to produce two copies of itself. The two copies replicate to produce four, and so on. During replication, miscopying occurs, and the erroneous copies that result are eliminated by selection. Only perfect copies survive. It is clear that, after each copying, at least one copy on average must be perfect. Otherwise selection cannot maintain the integrity of the message. This places an upper limit on the permissible mutation rate per base copied, or, equivalently, an upper limit on the length of the message, for a given mutation rate.

If the genome size, or the mutation rate per symbol, rises above this critical upper limit, the result is an accumulation of mutated messages. This is what Manfred Eigen and Peter Schuster have called the 'error threshold'. It is easy to see roughly where this upper limit lies. The requirement is that at least one perfect copy, on average, must be made at each replication. If there are n symbols, this means, approximately, that the probability of an error when replicating a symbol must be not greater than $1/n$. In other words, if the genome contains 1000 bases, the mutation rate per base, per replication, must be not greater than 1/1000.

The error rate in experiments of the kind illustrated in Fig. 3.1 is in the range 1/1000 to 1/10 000. This would permit a genome between 1000 and 10 000 bases. But this involves replication by an enzyme; if there is no enzyme, the error rate is much higher. Figure 3.2 shows an experiment by Leslie Orgel. In this experiment, an 'error' consists of a base other than C pairing with a G in the template. The error rate depends on the medium, the temperature, and so on, but very roughly the wrong base pairs with a G once in 20 times. This implies that, before there were specific enzymes, the maximum size of the genome was about 20 bases.

At first sight, this is a serious difficulty, and so it was long regarded. It presented a kind of catch-22 of the origin of life. Without a specific enzyme, the

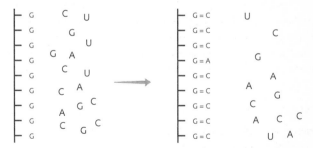

Figure 3.2 An experiment in base pairing. An RNA strand consisting entirely of a string of Gs is placed in a solution containing the four bases that constitute RNA: that is, A,C,G, and U. As expected, Cs pair with the Gs, but an occasional error occurs: in the figure, an A is shown pairing with one of the Gs. This pairing does not depend on the presence of any specific enzyme. The Cs, however, are not linked together to form a new strand: this does require an enzyme.

genome size is limited to about 20 bases; but with a mere 20 bases one cannot code for an enzyme, let alone the translating machinery needed to convert the base sequence into a specific protein.

Escape from this dilemma came from an unexpected source. It turns out that one does not need to have proteins and translating machinery to have enzymes. RNA molecules can themselves be enzymes. The significance of this discovery is discussed in the next chapter. In essence, the first RNA molecules did not need a protein polymerase to replicate them; they replicated themselves. But before we leave the problem of replication accuracy, there is another question to answer. We said that, in the presence of an enzyme, the error rate can be reduced to 1/1000–1/10000. This would suggest an upper limit to genome size of about 10000 bases. Yet, even when 'nonsense DNA' has been discounted, animals and plants have genomes of 10^9 to 10^{10} bases. How can that be?

In producing this book, the compositor will no doubt make some typographical errors. A proof copy will then be sent out, and we will find most of the mistakes and ask that they be corrected. No amount of proof-reading will reduce the errors in the final version to zero, if only because new errors can happen when correcting old ones; but the number of mistakes will, we hope, be less than it was in the first version. Exactly the same is true of DNA replication in higher organisms (bacteria and upwards). After the first enzyme-catalysed replication, there are in fact two error-correcting stages, called, reasonably, proof-reading and mismatch repair. This is possible because each DNA copy consists of an original strand and a new, copied strand. Enzymes in the cell check the base pairing between new and old strands, and if there is a mismatch—that is, a non-complementary pair—alter the new strand to match the old one. After two proof-reading stages, the error rate is reduced to one in 10^9 or less. Biologists are still arguing about whether this is as low as natural selection can make it without excessive cost, or whether it is an optimal compromise between the need to reduce the frequency of harmful mutations, and the desirability of producing an occasional good one.

CHAPTER 4

FROM THE RNA WORLD TO THE MODERN WORLD

In the last chapter, we explained an apparent 'catch-22' in the origin of life. Complex protein enzymes cannot exist without substantial lengths of DNA to code for them, but only rather short pieces of DNA could be reliably copied without enzymes: in short, no enzymes without long DNA molecules, and no long DNA molecules without enzymes. An escape from this catch-22 of the origin of life came with the discovery that RNA molecules can act not only as templates for copying but also as enzymes.

The RNA world

First, a word must be said about the differences between RNA and DNA. The chemical differences are minor; the chemistry of the backbone is slightly different, and one of the bases of DNA, thymidine, is replaced in RNA by uridine. In base pairing, however, uridine can replace thymidine as a pair for adenine, so that information can be passed from DNA to RNA, in a process called transcription (and, less commonly, from RNA to DNA, in 'reverse transcription'). These differences need not bother us.

A more important difference is that DNA usually exists in the cell as two complementary paired strands, in the famous double helix, whereas RNA usually exists as a single strand. This difference has two consequences. First, the proof-reading processes described on p. 36 do not occur in RNA replication, because the new and old strands separate immediately, and it is therefore impossible to check that base pairing is correct. In consequence, the error rate in RNA replication is in the range 1/1000 to 1/10000. RNA is therefore suitable as the genetic material only in organisms with very small genomes—that is, in viruses.

The second consequence of the single-stranded nature of RNA is that molecules of RNA can have a variety of secondary structures, depending on their sequences. Figure 4.1 shows an RNA molecule that was the end point of evolution in a test-tube, as described on p. 34. The molecule is bent back on itself, in several hairpin loops. The stems of the loops are maintained by base

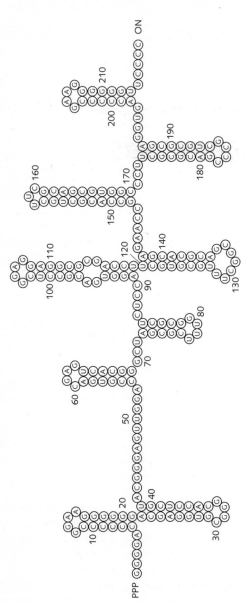

Figure 4.1 An RNA molecule that evolved in a test-tube.

pairing, and therefore their positions in the molecule are determined by the base sequence. Notice that pairing is always between strands in reverse orientation. DNA and RNA strands have a polarity, and pairing can occur only between strands pointing in opposite directions. Figure 4.1 shows only what is called the 'secondary structure' of the molecule; that is, the positions of the hairpin loops. A further folding process then leads to a three-dimensional tertiary structure. RNA molecules, therefore, can have a diversity of three-dimensional structures, whereas all DNA molecules have much the same double-helical form.

The point of all this is that RNA molecules can have a three-dimensional structure determined by their base sequence, just as protein molecules have a three-dimensional structure determined by their amino acid sequence. This led Carl Woese, Francis Crick, and Leslie Orgel to suggest, in 1967, that RNA molecules, like proteins, should act as enzymes. Since then, a rather rich list of RNA enzymes has been discovered, starting with the discoveries of Thomas Cech and Sydney Altman in the early 1980s, for which they received the Nobel prize for chemistry in 1989. The first ribozymes, as RNA enzymes are called, to be discovered in living organisms operate on nucleic acids as substrates, and use base pairing as a means of interacting with their substrate. Experiments are now in progress, using modified versions of the 'evolution in a test-tube' experiment shown in Fig. 3.1, to discover how wide a range of enzymatic activities are possible for ribozymes.

The relevance of ribozymes for the origin of life is enormous. As first argued by Walter Gilbert, instead of the existing division of labour between nucleic acids as carriers of information and proteins as enzymes, we can imagine a world in which RNA molecules performed both functions. A complex biochemistry could have evolved in this RNA world. Nevertheless, the problem of how the division of labour later came about still remains. Essentially, this is the problem of the origin of the genetic code, whereby base sequences determine amino acid sequences. The problem, although still difficult, is much easier if we suppose, as the discovery of ribozymes allows, that, even before the origin of the code, there were ribozymes able to catalyse a range of chemical reactions. Below, we offer a scenario for the origin of the code, but first we will describe how translation works today. Those already familiar with the elements of molecular biology should skip the next section.

How translation works today

The essential features of the translating machinery are shown in Fig. 4.2. Below is shown a 'messenger RNA' molecule, or mRNA for short. The base sequence of this molecule carries the information specifying a protein; it acquired this

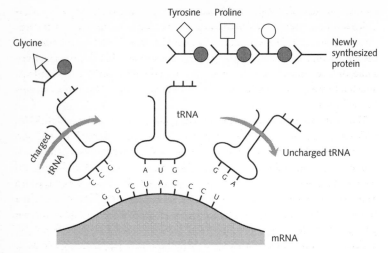

Figure 4.2 The translation process. The base sequence in the messenger RNA (mRNA) was acquired by transcription of a DNA gene in the nucleus. A triplet, UAC, of the mRNA is shown pairing with the 'anticodon' of a transfer RNA (tRNA) molecule to which is attached the amino acid tyrosine. Hence tyrosine is now added to the growing protein. This is what is meant by saying that UAC codes for tyrosine. This happens on the surface of a ribosome, which is not shown in the diagram. The ribosome holds the reacting molecules in the correct orientation, and forms the peptide bonds linking the amino acids together. The last tRNA to be used inserted a proline: it is shown on the right, leaving the ribosome. The next tRNA, carrying glycine, is shown on the left, just arriving: its anticodon will pair with the next codon, GGC. (There is an important sense in which this diagram is misleading. It shows only one tRNA, the correct one, just arriving. In practice, many different tRNAs would be jostling for position, but only the correct one would be able to pair with the anticodon and deliver its amino acid. It is as if 20 cars were competing for the same parking place: they drive into the slot in turn, but unless they have the 'right' anticodon they will be rejected, and be forced to back out and let someone else try. It is therefore not surprising that protein synthesis is the rate-limiting step in the growth of living organisms.)

sequence by the standard complementary base pairing procedure from one strand of a DNA gene. The mRNA passes to a structure called a ribosome, where it is translated; you can think of the ribosome as a tape recorder. The figure shows one new amino acid being added to the growing chain. More precisely, what is happening in the figure is that a triplet of bases, UAC, in the mRNA is being 'translated' as the amino acid tyrosine. How does this happen? The appropriate amino acid, tyrosine in this case, is already attached to a 'transfer RNA molecule', or tRNA. The tRNA molecule has a loop, on which are exposed three bases (AUG in the figure), forming an 'anticodon'. This anticodon pairs, in a complementary manner, with a triplet of bases in the mRNA (UAC in the figure; remember that A pairs with U and G pairs with C). This triplet, UAC, is called a codon. This particular codon is said to 'code for' the amino acid tyrosine. This coding assignment is brought about because the tRNA molecule with the anticodon AUG also has attached to it the amino acid tyrosine. The ribosome, or tape recorder, has enzymes that detach the amino acid from the tRNA, attach it to the growing amino acid chain, and release the tRNA, which can now be used again.

The figure shows the last tRNA molecule, now released, and the next tRNA, with the anticodon CCG, ready to pair with the next codon, GGC, and also bearing the amino acid glycine, coded for by GGC. This next pairing, however, cannot happen until the UAC codon has been decoded, the tRNA released, and the mRNA moved along the ribosome so that the GGC codon is exposed in a 'slot'. Note also that when the GGC codon is so exposed, it will usually not be a tRNA with the appropriate anticodon, CCG, that first arrives. Collisions will take place randomly until the right tRNA does arrive. Imagine the tRNA molecules as motor cars, with symbols on the bonnet. They must take it in turns to drive into a parking place. A car must back out again if its symbol turns out to be wrong: only when, by chance, it has the right symbol can it deliver the load it is carrying. Successful delivery would be a slow process. The analogy is a fair one, so it is not surprising that this process of protein synthesis turns out to be the one that, when everything is favourable, limits the rate at which living cells can grow.

The association between amino acids and codons—for example, between UAC and tyrosine—is called the genetic code. Three of the possible 64 codons do not code for an amino acid, but are 'stop' codons, terminating translation. When such a stop is reached, no further amino acids are added to the chain, which is released from the ribosome as a completed protein. The remaining 61 codons specify 20 different amino acids, so the code is redundant: that is, most amino acids are coded for by more than one codon.

It will be clear that the meaning of the code—that UAC specifies tyrosine and not some other amino acid—depends on the fact that the tRNA with the anticodon AUG also has attached to it the amino acid tyrosine. In existing tRNAs, the site of attachment of the amino acid is distant from the anticodon. The attachment of the appropriate amino acid is carried out by a specific enzyme, which we will call an assignment enzyme, because, in effect, each such enzyme assigns a particular codon to a particular amino acid. The code is therefore chemically arbitrary. By altering the sequence of a tRNA, or the specificity of an assignment enzyme, the code would be altered. Mutations, usually lethal, that alter the code in these ways are known. It is still an open question whether the code was always arbitrary or whether there was once a good chemical reason why UAC specified tyrosine.

All existing organisms have essentially this system. Viruses, which can grow only inside cells, do not themselves code for a translating system (that is, ribosomes, tRNAs, and assignment enzymes) but rely on the translating system of the cells they infect to synthesize the proteins for which they code. We have, therefore, no intermediate between a complete system and no system at all to guide us in guessing at its evolution. We can, however, identify one central characteristic that needs to be explained. This is the attachment of a particular amino acid to a particular sequence of RNA bases. It is this characteristic that is the clue to the origin of the code.

The origin and nature of the code

In evolution, complex organs adapted for particular functions often arise in a simpler form, with a different function. A classic example is the idea that feathers first evolved as modified, frilly scales, serving to preserve heat, as down feathers do today, and only later acquired the specialized shapes and web-like structure of flight feathers. We think that something of this kind happened in the origin of the code. As explained in the last section, the crucial step was the chemical bonding of particular amino acids to small RNA molecules with specific base sequences. Our suggestion is that this first happened not as part of a protein-synthesizing apparatus but to improve the range and efficiency of ribozymes.

The idea is that ribozymes acquired amino acids as 'cofactors': that is, an amino acid was attached to a ribozyme and made it a more efficient catalyst. Today, protein enzymes often have cofactors (not, of course, amino acids, which they already have), bound to the surface of the enzyme at the active site and participating in the reaction. By using cofactors, the range and specificity of catalytic activity can be increased. Ribozymes would have been in greater need of cofactors than protein enzymes, because the range of reactions they can catalyse without them is probably much smaller—they are composed of only four chemically different monomers instead of 20. Amino acids would have been the obvious candidates because, as Stanley Miller's experiments showed, they would have been abundant on the primitive Earth.

One way of attaching an amino acid to a particular point on the surface of a ribozyme is by complementary base pairing, as shown in Fig. 4.3. The amino acid is bound to an oligonucleotide (that is, an RNA molecule consisting of only a few nucleotides—three in the figure), and the oligonucleotide is bound to the surface of the ribozyme by base pairing. The ribozyme in question—let us call it R_1—catalyses some chemical reaction. In modern organisms, that same reaction is catalysed by a protein enzyme. It is not reasonable to suppose that the protein enzyme evolved quite independently of the ribozyme, until it was ready to

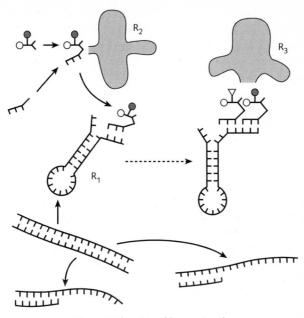

Figure 4.3 The origin of the genetic code.

take over, and that the ribozyme then was lost. It is more reasonable to imagine, as suggested by H. B. White in the 1980s, that the ribozyme was gradually transformed into a protein enzyme, via a series of intermediate hybrid enzymes. Initially, the enzyme was composed of RNA, perhaps with an amino acid cofactor: today, it is composed of amino acids, perhaps with a nucleotide cofactor representing a trace of its RNA origin.

We have, then, two basic ideas. One is that the first linking of an amino acid to an oligonucleotide occurred in the construction of cofactors; the second is that ribozymes were gradually transformed into protein enzymes. Figure 4.3 shows how these ideas can be the basis of a scenario for the origin of the code. The first point is that an amino acid would not be attached to an oligonucleotide by accident; the attachment would have to be catalysed by an enzyme, R_2 in the figure. Given the existence of the ribozyme R_2, many identical cofactors could be synthesized, each consisting of an oligonucleotide attached to an amino acid, and could be used by different ribozymes. Initially only one kind of amino acid and one kind of oligonucleotide would be used. Later, a second amino acid, attached to a different oligonucleotide by a different ribozyme, would be added, and so on.

As explained in the last section, the crucial feature of the genetic code is the attachment of particular amino acids to tRNA molecules, a step carried out by assignment enzymes. We are suggesting that this attachment first occurred to make cofactors, and was carried out by ribozymes. The implication of Fig. 4.3 is that these ribozymes—for example, R_2—were the ancestors of today's assignment enzymes, and that the oligonucleotide components of the cofactors were the ancestors of today's tRNA molecules.

How could such a system evolve into today's protein-synthesizing machinery? We can only answer this question in vague outline, as shown in Fig. 4.3. The first step, presumably, was the use of two or more amino acid cofactors by a single ribozyme. If these amino acids were linked together by peptide bonds, we are on the way to forming a protein enzyme. In the figure, we have shown the peptide bond being formed by a third ribozyme, R_3. We can now see, albeit hazily, the ancestors of the various components of the modern translating machinery. As we have already said, ribozymes like R_2 were the ancestors of assignment enzymes, and the oligonucleotide part of the cofactors were the ancestors of tRNA. The ribozyme R_1, originally the ribozyme that acquired cofactors, would in time transfer its original function to the peptide bound to its surface, and evolve into the messenger RNA against which the tRNAs line up. Finally, the ribozyme R_3, forming the peptide bond, is the first rudiment of today's ribosome. It is intriguing that the ribosomal enzyme that today catalyses the formation of peptide bonds is probably a ribozyme.

This outline is necessarily speculative. But it does have the virtue of suggesting a series of functional intermediates between the RNA world and the present system. Even one cofactor consisting of an amino acid linked to an oligonucleotide would be worth having, and new amino acids could be added one by one. The model also suggests how a ribozyme could gradually be replaced by a protein enzyme, without requiring that the protein evolved independently until it was ready to take over.

We have suggested how the genetic code may have evolved, but have said nothing about why the code is the particular one that we see. Why do particular triplets stand for the amino acids they do; for example, why does UUU code for phenylalanine? The first point to make is that the code is almost universal. There are a few variations; for example, AAA, which specifies lysine in the universal code, specifies asparagine in the mitochondria of flatworms and echinoderms. This is puzzling because, once the code has been established, it is hard to see how it can evolve. Thus imagine that the assignment enzyme that attaches lysine to the tRNA complementary to AAA were to mutate, so as to attach asparagine instead. This would cause asparagine to be inserted in place of lysine in many different proteins. Surely some of these changes would be lethal or harmful. The difficulty is not insoluble. The explanation depends on the fact that the code is redundant. Lysine is coded for not only by AAA but also by

AAG. If, in some lineage, all lysines were coded for by AAG, and the codon AAA came not to be used, then AAA could be reassigned to code for asparagine; that is, a mutation causing asparagine to be attached to the complementary tRNA could occur without harmful effects.

Some changes of this kind have happened, but they are few. The code is so nearly universal that it is strong evidence that all life on Earth had a single origin. The only alternative is that the codon assignments are necessarily as they are; that is, there is some chemical reason why AAA must code for lysine, AUG for methionine, and so on. There is still debate about whether there is any chemical sense in the particular assignments observed, but few people would argue that any sense there is amounts to necessity. Universality implies a common origin.

Although some features of the code may reflect the accidents of origin, there are others that appear adaptive, and so suggest that codon assignments were influenced by selection. In particular:

1. Similar codons specify the same amino acid; for example, GUU, GUC, GUA, and GUG all specify valine.
2. Similar codons specify chemically similar amino acids; GAU and GAC specify aspartic acid, and GAA and GAG specify the chemically similar glutamic acid.
3. Amino acids that are used more often in proteins are specified by a greater number of different codons; for example, the commonest amino acid, leucine, is coded for by six codons, and the relatively rare tryptophan by only one.

The first two features have the effect of reducing the harmful effects of mutation, because many mutations will either not alter the protein at all or will replace an amino acid by a chemically similar one. How could such an apparently adaptive feature evolve if, as we have argued, the code changes very rarely? Natural selection cannot adapt something that does not vary.

The answer may be that these apparently adaptive features of the code originated, not by natural selection between different codes, but in other ways. When it first evolved, the code was less specific than it is now. Groups of similar codons may have specified any one of a group of similar amino acids, because the assignment enzyme could not tell them apart. If, starting from such a rather fuzzy set of assignments, the code gradually became more precise, this would account for the first two features. The third feature looks as if it might be an adaptation for more efficient synthesis. But it may have an entirely non-adaptive explanation. Once the code was fixed, random mutations to a codon specifying leucine were six times as common as mutations to the single codon specifying tryptophan. If at least some amino acid changes occur because they do not matter, this would explain why leucine is commoner in proteins than tryptophan. Thus the truth may be, not that there are more codons for leucine because leucine is more often needed, but that there is more leucine in proteins because there are more leucine codons. The truth may lie somewhere in between.

The nature of the genetic code is one of the most astonishing things we know about the world. It parallels in importance our knowledge of the structure of the Solar System or Dmitri Mendeleyev's table of the chemical elements. To explain its origin and evolution is a major challenge, which is just beginning to be met.

CHAPTER 5

FROM HEREDITY TO SIMPLE CELLS

The scenario we outlined in the last chapter for the origin of the code assumed co-operative interactions between replicators. For example, one ribozyme functioned to produce a cofactor that would increase the efficiency of a second ribozyme. Even if this particular hypothesis turns out to be mistaken, it is clear that, for life to progress beyond the stage of simple replicating molecules to something more complex, a web of co-operative interactions between genes had to evolve. In this chapter we discuss how this could have happened.

From ecology to individuality

Imagine, as a starting-point, a system consisting of several different kinds of replicating molecules. These would have interacted in various ways. Most obviously, they would have competed for a limited supply of the monomers of which they were composed. If so, the molecules would resemble competing organisms in an ecosystem. Other kinds of interaction are possible. One type of molecule might aid the replication of a second, which in turn aided the replication of the first: this would be a case of a pair of molecular mutualists. Or a molecule could, by virtue of its enzymatic function, attack a different type and cut it into pieces so that it could not replicate, and then use the building blocks for its own replication: this would be an example of predation. Finally, a molecule might use the help of another in its own replication, and do nothing in return: such a molecule would be a parasite. In effect, the replicating molecules would form a simple ecosystem.

This parallel between the first populations of replicators and an ecosystem needs emphasizing. Consider a present-day ecosystem—for example, a forest or a lake. The individual organisms of each species are replicators; each reproduces its kind. There are interactions between individuals, both within and between species, affecting their chances of survival and reproduction. There is a massive amount of information in the system, but it is information specific to individuals. There is no additional information concerned with regulating the system as a whole. It is therefore misleading to think of an ecosystem as a super-organism. In particular, it would be wrong to expect individual organisms, or species, to have properties that evolved to ensure the survival of the whole ecosystem, as the parts of an organism—for example, its eyes or its leaves—evolved to ensure the survival of the individual. This is because there has been no natural selection favouring the survival of one ecosystem at the expense of another. The point should be obvious, but it is sometimes not appreciated. For example, the Gaia hypothesis, which likens the Earth and the biosphere to an organism, is useful if it leads people to value the biosphere, and scientifically correct in that it emphasizes the role of living organisms in altering the climate and the chemical composition of the Earth's oceans and atmosphere. But it is misleading if it leads people to expect the Earth to defend herself, as an animal or plant might defend itself.

What is true of present-day ecosystems was true of primitive ecosystems composed of replicating molecules. There was no general reason why particular replicators should evolve characteristics favouring the survival of the whole, and no reason to expect co-operative rather than competitive interactions. Of course, some co-operative interactions would arise by chance. Manfred Eigen and Peter Schuster have argued for the importance of one particular type of co-operative network, which they called the hypercycle, illustrated in Fig. 5.1.There are several kinds of replicator, A, B, C, and D. The rate of replication of each kind is an increasing function of the concentration of the replicator immediately preceding it in the cycle. Thus the rate of replication of B increases with the concentration of A, and so on round the cycle: the cycle is closed, because the rate of replication of A increases with the concentration of D.

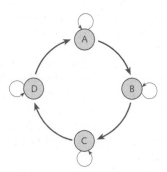

Figure 5.1 The hypercycle. Each of the units A, B, C, and D is a replicator. The rate of replication of each unit is an increasing function of the concentration of the unit immediately preceding it. Thus the rate of replication of B is an increasing function of the concentration of A, and so on round the cycle.

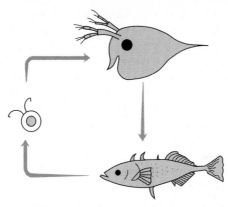

Figure 5.2 An ecological hypercycle. The three replicators are *Daphnia* (a waterflea), *Chlamydomonas* (a single-celled green alga), and the stickleback. *Daphnia* feed on algae, and reproduce faster if there is plenty to eat: the same is true of sticklebacks, that feed on *Daphnia*. The rate of growth of the algae will depend, among other things, on the amount of nitrate in the water, and sticklebacks excrete nitrate. Thus each member of the cycle encourages the reproduction of the next one.

The interest of the hypercycle is that it is ecologically stable, even if the four kinds of replicator compete for the same resource—for example, if they are composed of the same monomers. In contrast, if there were no interactions of the kind shown in Fig. 5.1, it is inevitable that one of the kinds would win in the competition for resources, so that only one kind would survive. The system illustrated in Fig. 5.1 may seem very abstract, but in fact ecosystems are full of hypercycles. One such is illustrated in Fig. 5.2. Algae, *Daphnia*, and sticklebacks are all replicators. The rate of replication of *Daphnia* increases with the concentration of algae, and of sticklebacks with the number of *Daphnia*. The cycle is closed because fish excrete into the water nitrogenous compounds which accelerate the growth of algae.

Hypercycles can help to stabilize ecosystems and to encourage growth. This would have been as true of a system composed of replicating molecules as of a modern ecosystem. But we cannot expect a hypercycle to evolve increasing cooperation between its components. Indeed, it is just as likely that evolution will disrupt the cycle. For example, we can expect evolutionary changes in *Daphnia* that make it better at eating algae, but not changes that make *Daphnia* easier for fish to eat. In fact, *Daphnia* have evolved the ability to respond to the presence of fish by growing spines that make them harder to eat. In the case of a molecular cycle, if we imagine each replicator as a 'replicase' acting to copy the next member of the cycle, then there are two kinds of mutation that would improve the cycle: mutations making a molecule a better replicase, and mutations making it a better target for replication. Selection will favour the latter type of mutation but not the former.

There is, however, one circumstance in which both types of mutation might be favoured by selection. Suppose that all molecules are enclosed within 'compartments', or cells. If we assume that cells in which the total number of replicating molecules increases most rapidly also divide most rapidly, then cells containing the most rapidly replicating hypercycles would increase in frequency relative to others. In effect, by enclosing groups of molecules within cells, we have introduced a new level of selection: there is now between-cell selection as well as selection between molecules within a cell. In the next section we consider the effect of enclosing populations of molecules within cells. After that, we consider how cells might have arisen.

Why cells? The stochastic corrector model

What do we require of an early cell? It should be enclosed in a bag, or membrane, which should grow with the cell. The membrane should permit nutrients to pass through, but be impermeable to macromolecules and to the smaller molecules involved in metabolism. The result of metabolism must be the supply of building blocks for new membrane and new genetic material, as well as additional components of the metabolic network itself. Finally, the processes must be under the control of the genetic material which, as a legacy of the precellular phase of evolution, was a collection of genes not physically linked to one another. The genes are likely to have been RNA-like molecules acting as catalysts of various cellular processes, particularly those of metabolism. Referring back to Chapter 1, the chemoton there described is an adequate model of this stage of evolution. Although we are still far from the artificial synthesis of such a system in the laboratory, this may happen in a few decades.

The model we now describe—the stochastic corrector model—has been studied by computer simulation and by mathematical analysis. But the model itself is simple, and its behaviour can be understood without recourse to mathematics (Fig. 5.3). It makes the following assumptions:

1. There are two kinds of replicator, A and B. One kind, A, replicates faster than the other, so that, if there were no cells, B would be eliminated by selection.

2. The replicators are enclosed within cells, with only a few molecules in each cell. When the number of molecules in a cell rises to some critical value, the cell divides into two. At division, the molecules are distributed randomly to the two daughter cells. If a daughter cell contains no replicators, it dies: otherwise, replication continues.

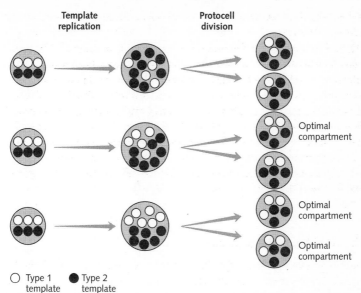

○ Type 1   ● Type 2
  template   template

Figure 5.3 The stochastic corrector model. The initial compartments each contain three copies of type 1 templates, and three copies of type 2 templates. During replication, there is a tendency for 'black' templates to outgrow 'white' ones, but, because of the small numbers of templates, there are chance variations as well. It is also assumed that protocells with both kinds of template grow faster than cells with only one kind. When the protocells divide, templates are distributed randomly to the daughter cells. Mathematical analysis shows that the population settles down to an equilibrium, with a constant proportion of optimal protocells, with three templates of each kind. In this model, there is 'group selection' favouring protocells with both kinds of template, and 'individual selection' favouring 'black' templates within each cell. Because of the small numbers of templates per cell, and the resulting chance variation, group selection can win out.

3. Although, within a cell, A molecules replicate faster than B, the rate of replication of both types is greatest if A and B molecules are present in equal numbers, and slowest if one type is completely absent. Thus there is a synergistic interaction between the two kinds, or, in other words, A and B cooperate.

Given these assumptions, it turns out that cells containing both kinds of molecule, A and B, survive, although from time to time an A-only cell, or a B-only cell, will arise by chance. This result—the survival of co-operators—depends on there being rather few molecules per cell. The reason is as follows. If the numbers are small, chance events, and in particular the chance segregation of molecules when the cell divides, produces differences between cells in the numbers of molecules that they contain, and so makes selection between cells effective. It is because of the importance of chance events that the model is called the 'stochastic' corrector.

We have described the simplest kind of model, with only two kinds of replicator, but any type of co-operative system could be favoured once there were cells. Thus efficient hypercycles would evolve, but so would any other type of co-operative system. There is, however, a limit to what the model will do. We have emphasized that the total number of replicators per cell must be small. It follows that only rather simple systems, with a small number of different kinds of replicator, can be maintained. A system depending on a large number of different kinds would soon collapse, because of the risk of losing one of the kinds completely when the cell divides. In modern cells, this problem is solved by linking the genes together on chromosomes.

The origin of chromosomes

The linkage of genes on chromosomes has two advantages. First, it ensures that when one gene is replicated, all are: no one can cheat. Second, it makes it easier to arrange that, when the cell divides, one copy of each gene passes to each daughter cell. These two advantages were not achieved simultaneously. Linkage, ensuring synchronous replication, evolved before regular segregation at cell division. The origin of linkage can be studied by using an extension of the stochastic corrector model. Consider, as before, a case in which there are only two kinds of replicator, A and B. Imagine that a new joint molecule, with A and B linked end to end, arose as a rare mutation. Would such an 'Ur-chromosome' spread by natural selection? The snag is that, being twice as long, the A–B molecule would replicate more slowly than single A and B molecules, so that selection within cells would tend to eliminate the new A–B molecules. Yet computer simulation shows that A–B chromosomes can spread, provided that the same conditions hold as are needed for the coexistence of the two kinds of single molecules. There must be few molecules per cell, and the two kinds must co-operate: that is, cells containing both kinds must grow faster than cells containing only one kind.

There are two ways of seeing intuitively why A–B chromosomes succeed, without doing the simulations. One is to note that any cell with a chromosome does in fact contain both kinds of function, and therefore grows faster: there is between-cell selection favouring the chromosome, which can counterbalance the within-cell selection favouring separate replicators. The other way is to take a gene's-eye view. An A gene can either be single, in which case it replicates faster within the cell, or it can be linked to a B gene, in which case it replicates more slowly but is sure to be in a 'fit' cell , containing both A and B functions, after the cell divides. Provided that the number of replicators per cell is small, and the advantages of co-operation substantial, it turns out that chromosomes can spread.

If two linked genes can replace single genes, then additional genes, C, D, and so on, can be added to the linkage group. This can happen even though there is no segregation mechanism ensuring the passage of one chromosome to each daughter cell. In Chapter 6 we describe the very different segregation mechanisms found today in the two main types of cell, prokaryotes and eukaryotes, and discuss how the latter could have evolved from the former. But even the prokaryote mechanism is too complex to be primitive. We have at present no clue as to the nature of the earliest segregation mechanism.

Membranes and cells

We have been supposing that populations of replicating molecules came to be enclosed within membranes. How could this happen? As usual, we first describe how things are done today, and then ask about origins.

Existing membranes (Fig. 5.4) consist of a lipid bilayer, formed of long-chain fatty acids. This sounds complicated, but isn't. A fatty acid is a linear molecule with two chemically different ends. One end is 'hydrophobic': that is, like fat or oil, it will not mix with water. The other end—the acid end—is hydrophilic: it mixes readily with water. If fatty acid molecules are shaken up in water, lipid bilayers are formed spontaneously. The hydrophobic ends avoid the water, and join one another to form the centre of the bilayer. The hydrophilic ends form the outer surfaces of the bilayer, in contact with the water. Not only bilayers but spherical vesicles form spontaneously. This happens because, if there are flat sheets, the hydrophobic ends are still in contact with water at the edges of the sheets. Only by folding into a sphere can all contact between hydrophobic molecules and water be avoided.

At first sight, then, it might seem that the origin of cells is a non-problem. Unfortunately, this is not so. There are three difficulties. The first has already been mentioned. Where did the long-chain fatty acids come from? They are not formed in Miller-type primitive-soup experiments. Our hope is that their formation on charged surfaces—in a primitive pizza rather than a primitive soup—will prove more feasible, but experimental investigation of the process is still needed. The other two difficulties concern the transport of substances into and out of primitive vesicles, and the mechanism of division.

In existing organisms, there are complex proteins embedded in the cell membrane that allow molecules to enter and leave. These are of two main kinds: proteins with small holes, or pores, that permit the passage of small molecules, and so-called 'permeases' that actively pump specific molecules across the mem-

(i)

(ii)

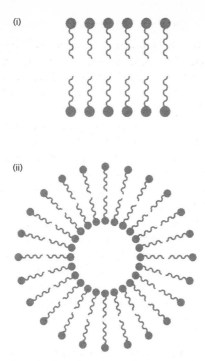

Figure 5.4 Structures formed by long-chain fatty acids in water. Each molecule has a 'head' that mixes happily with water, and a 'tail' that shuns water. In (i), they form a bilayered membrane, with the heads in contact with water, and the tails in contact with one another, avoiding water except at the edges of the sheet. The tails can avoid water completely by forming a vesicle (ii). Note that there is water inside the vesicle.

brane against a concentration gradient, using energy to do so. Clearly, such proteins could not have been primitive. In the absence of such proteins, only a few kinds of inorganic molecule—for example, carbon dioxide and hydrogen sulphide—would have been able to enter. Many essential molecules—in particular, charged molecules or ions, such as the phosphate ion—would have been unable to do so.

First, why did the first cell membranes have to permit the passage of small molecules? After all, was not the whole point of the first membranes to prevent the entry or escape of replicators? Indeed it was, but these same membranes must have permitted some small molecules to pass. To explain why, we must say something about where these first cells got their energy from. As we have already emphasized, cells are chemical engines, and engines must be supplied with energy.

Today, cells get their energy in one of three ways:

1. *Photosynthesis*. Both the prokaryotic cyanobacteria (blue-green algae) and the eukaryotic algae and higher plants get their energy direct from sunlight. They are able to trap the energy in sunlight and use it to synthesize energy-rich organic compounds, in particular sugars. These compounds are then used as fuel to drive the rest of the cell's metabolism.

2. *Heterotrophy*. Animals and fungi, and non-photosynthetic cells generally, take in energy-rich compounds as food. The ultimate source of such compounds is photosynthesis. Animals eat plants, or they eat animals that have eaten plants. Fungi and many bacteria live on decaying animals and plants. (The root 'hetero' here means that an organism is relying on another organism for its source of energy.)

3. *Autotrophy*. Some bacteria synthesize their organic material exclusively from compounds with only one carbon atom, such as carbon dioxide. The energy for this synthesis can be obtained either from sunlight (photosynthesis) or from inorganic sources. For example, hydrogen sulphide is present in hydrothermal vents in the deep sea and in volcanic outflows; energy can be obtained by oxidizing it. Today, the ecosystems of deep-sea vents rely largely on this reaction, rather than on photosynthesis, as their ultimate source of energy.

What of the first cells? Clearly, photosynthesis is not a plausible answer, because it requires complex and highly evolved proteins. By definition, the first organisms must have been autotrophic: they could not have relied on other organisms for their energy. But the primitive ocean may have been rich in abiotically synthesized organic compounds, similar to those synthesized in Stanley Miller's experiments. The source of the energy in these compounds was sunlight. The efficiency of their formation—that is, the energy in the compounds formed, as a fraction of the energy in the sunlight—would have been very low compared with the efficiency of photosynthesis, but that would not have mattered. The snag is that it is hard to see how this supply of energy could have entered the first cells. Sugars are not going to pass through a lipid bilayer, without the help of specialized protein pumps. It seems that we are left with autotrophy, based on reactions between inorganic compounds, as the source of energy in the first cells.

A possible solution to the transport problem is that true cells were preceded by 'semicells' that formed as minute blisters on mineral surfaces (Fig. 5.5). There are good chemical reasons why fatty acids may have been first synthesized on such surfaces, and a concentration of fatty acids could lead to the formation of semicells. Such semicells would have access to a range of molecules that were either components of the mineral itself, or were attached to its surface. This

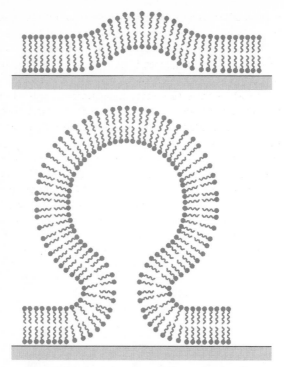

Figure 5.5 The formation of a semicell on a mineral surface.

would provide it, not only with the elements from which its own component molecules could be synthesized, but also with a source of energy. During such a semicellular stage, the permeability of the membrane to desirable compounds present in the surrounding water could have been gradually increased, by inserting into the membrane molecules that would permit their passage. As always when explaining the evolution of novelty, we seek an intermediate stage that would work well enough to survive, giving an opportunity for gradual improvement by natural selection. The semicell, relying on mineral surfaces for nutrients, provides such an intermediate.

There remains the problem of how the first cells could divide. The problem may not be as difficult as it seems at first sight. Imagine a spherical vesicle within which fatty acids (from which new membrane is formed) are synthesized at the same rate as other internal components such as replicators and ribozymes. As the fatty acids are made, they will automatically insert themselves into the membrane, which will grow in area. Starting from a spherical vesicle, suppose that all components, both of the membrane and internal constituents, increase by a factor of two. If the surface area was doubled, and the spherical form was retained, the increase in volume would be more than doubled, although the internal constituents would only have doubled in amount. What will happen? The natural outcome is that the membrane will buckle, and the vesicle will divide into two. This conclusion is beautifully confirmed by the experiments of Pier Luigi Luisi, who managed to combine a replicating template with a membrane subsystem. The microsphere divided while the genetic material replicated, retaining the relative proportions of the two components. This is already a big step towards a chemoton.

When the vesicle divides, its contents, including any replicators that may be present, will be partitioned randomly between the daughter vesicles. This is what we assumed earlier, when discussing the evolution of co-operation between replicators, and the origin of chromosomes. Only later would a mechanism evolve that ensured chromosome segregation, so that each daughter vesicle was guaranteed to receive a complete set of replicators, or genes. When this step had been taken, true cells would have come into existence.

CHAPTER 6

THE ORIGIN OF EUKARYOTIC CELLS

Our cells are much more complex than those of bacteria. We belong, with many other living species, to the large group (or empire, as taxonomists call this highest category) of eukaryotes. The name comes from the Greek, and refers to organisms with a 'proper' nucleus, the central body of the cell, separated by two membranes from the cytoplasm. Nuclei are easily seen through a light microscope. They harbour the chromosomes, in which the physically linked genes are packaged by proteins into coiled structures. In contrast, bacteria usually have only one chromosome, which lacks a similar coiled structure. Furthermore, the bacterial chromosome , although anchored to the cell wall, is free in the cytoplasm. Bacteria (including cyanobacteria, traditionally called blue-green algae) are therefore called prokaryotes: they lack a proper nucleus. Typically, eukaryotic cells are much larger than prokaryotic cells: the former have, on average, a 10 000-fold larger volume.

What else, apart from the possession of a nucleus, distinguishes eukaryotic from prokaryotic cells? There are many organelles ('little organs') in the former that do not occur in the latter. These organs include the mitochondria, which play a special part in energy metabolism: they have two membranes, and multiply within the cell by binary fission. Photosynthetic eukaryotes (plants, green and red algae, and a curious group, discussed below, called chromists) have, in addition to mitochondria, a somewhat similar organelle, the plastid. Plastids synthesize organic compounds from inorganic ones at the expense of captured light energy, using special photosynthetic pigments to trap the energy. Plastids also have two membranes around them, and reproduce by fission.

The peculiar structure and autonomous division of plastids and mitochondria prompted some scientists about a hundred years ago to suggest that they may be the descendants of formerly free-living bacteria with similar metabolic capacities. These ideas were not taken very seriously: indeed, they were often ridiculed. Later, after the development of genetics, it was discovered that some mutations affecting mitochondria and plastids do not seem to be located in the nucleus: rather, the genes are associated with the organelles themselves.

Molecular biology then made possible the discovery within the organelles of small circular chromosomes and small ribosomes, both similar to those of bacteria. In the early 1970s, Lynn Margulis forcefully revived the idea of the symbiotic origin of plastids and mitochondria. The whole biological community has been converted to this view by now.

Even if we accept the symbiotic theory of the origin of organelles, this does not explain the origin of the eukaryotic cell itself. First, there is the nucleus: where did it come from? Is it too descended from a former symbiont, as some have suggested, or did it, as we think is more likely, evolve without symbiosis? The eukaryotic cytoplasm has other structures not found in prokaryotes. In particular, there is a system of membranes known as the endoplasmic reticulum, whose origin and significance require explanation.

Perhaps the most significant difference between prokaryotic and eukaryotic cells, if we want to understand the origin of the latter, is that the former are enclosed in a rigid cell wall (or at least a very rigid cell membrane), whereas many eukaryotic cells have no cell wall, and change their shape vigorously. The absence of a cell wall makes possible a process called 'cytosis', illustrated in Fig. 6.1. A membrane-bounded vesicle within the cell can fuse with and become part of the cell membrane, or, in the reverse process, a vesicle can be formed from the outer cell membrane. Cytosis is involved in many cellular processes, but perhaps its original role was in 'phagocytosis' (Fig. 6.2), the process whereby a eukaryotic cell can feed on solid particles. Bacteria acquire nutrients by absorbing them molecule by molecule across the cell membrane. They can feed on solid objects only by first secreting digestive enzymes into the surrounding medium, and then absorbing the molecules produced by digestion—clearly a very wasteful process. Eukaryotes can engulf a food particle into a food vacuole, which can then fuse (by cytosis) with a lysosome containing digestive enzymes: in this way, no nutrients or enzymes are wasted. The invention of phagocytosis was probably responsible for the ecological success of the first eukaryotes.

If we accept that the loss of an outer cell wall and the evolution of phagocytosis

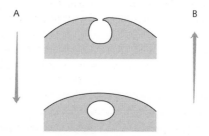

Figure 6.1 Cytosis. An inpushing of the cell membrane forms an intracellular vacuole (A) or a vacuole fuses with and becomes part of the cell membrane (B).

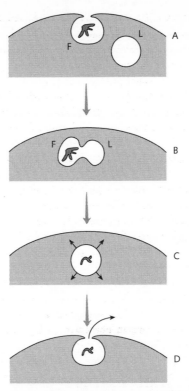

Figure 6.2 Phagocytosis. (A) A solid food particle is being engulfed into a newly formed food vacuole (F); a lysosome (L), containing digestive enzymes, has already been formed internally. (B) The food vacuole and lysosome fuse. (C) The particle is digested, and small molecules pass out of the vacuole into the cell cytoplasm. (D) The undigested remains of the particle are extruded from the cell.

were the initial steps in the origin of the eukaryotes, it is easy to see how this could have led to the evolution of symbiotic organelles. They could have arisen because ingested bacteria were not digested, a kind of 'cellular indigestion'. There is some support from the study of existing protists for the view that the acquisition of symbionts was secondary. The only known survivors of the most ancient eukaryotes are the so-called Archaezoa. They have a nucleus and rod-shaped chromosomes, but are never multicellular, and lack mitochondria and plastids. Their ribosomes resemble those of bacteria in that they are smaller than typical eukaryotic ribosomes. Molecular phylogeny confirms the view that present-day archaezoans are related to the first eukaryotes.

We now describe in some detail the transition from prokaryote to eukaryote.

Inevitably, the story is complicated, involving unfamiliar structures and unfamiliar technical terms. There is, however, a coherent thread that can be followed. The precipitating event seems to have been the loss of the rigid outer cell wall characteristic of bacteria. The reason for this loss is unknown, but, ever willing, we offer a speculation below. Having lost its 'external skeleton', however, the cell was forced to do several things to compensate. It had to invent a new 'internal skeleton' of filaments and microtubules. It also had to invent a new mechanism, called mitosis, whereby one copy of its chromosome (or chromosomes) was passed to each daughter cell at cell division, because the old, bacterial, mechanism of chromosome segregation depended on attachment of the chromosome to the cell wall, and so would no longer serve.

Despite the problems arising from the loss of the cell wall, there were compensating advantages and opportunities. As already explained, it made possible a new way of feeding, by swallowing solid objects: this was probably the immediate advantage that ensured the survival of the first eukaryotes. This new method of feeding led to the evolution of new intracellular structures, or 'organelles', derived from bacteria that were swallowed but not digested. The new method of chromosome segregation brought with it an unforeseen but profound advantage: the total amount of genetic information that a cell can contain, which is limited in bacteria because replication of the chromosome must start from a single point, could increase many-fold. We now look at these changes in more detail.

The catastrophic loss of the cell wall

The loss of the outer cell wall is important both in making possible a new way of feeding and also for other changes it brought in its train. It is now possible to piece together a more detailed history. Ordinary bacteria, as we have said, have a rigid cell wall. There is another group of prokaryotes, the so-called archaebacteria, that lack a cell wall but have a very rigid type of cell membrane instead. They live in esoteric environments, such as hot and acidic solfataras, strictly anaerobic places and so on, and have a variety of unusual metabolic systems. In 1977, Carl Woese discovered that the gene sequences of archaebacteria are very different from those of ordinary bacteria. Because they live in extreme environments, somewhat resembling those in which life may have begun, they were called 'archae'-bacteria.

At first, many researchers inclined to the view that the archaebacteria were indeed the most ancient living beings, direct descendants of the first living cells. Recent work, however, suggests that they are derived from ordinary bacteria (referred to as eubacteria). Already, before this was known, the British biologist Tom Cavalier-Smith had hypothesized that both archaebacteria and eukaryotes descended from eubacteria that suffered a catastrophic loss of the rigid cell wall. Why should this have happened? We cannot argue that a group of bacteria lost the cell wall because it would later enable them to evolve phagocytosis: this would require that evolution has foresight. A possible, though speculative, suggestion is that some bacterium evolved, as a competitive device, an antibiotic that blocked the synthesis of the cell wall of another bacterium: that is how some modern antibiotics (for example, penicillin) kill bacteria. However that may be, the suggestion is that it was the loss of the cell wall that triggered everything that was to follow.

Wall-less bacteria are extremely fragile, as laboratory experiments tell us. Such losses may have occurred several times in evolution. The most common outcome would have been the extinction of such lineages, unless they discovered an antidote to the poison. But two lineages, related to one another, found novel remedies to the disaster: one developed a rigid cell membrane using new types of membrane-forming molecules, and gave rise to the archaebacteria, whereas the other, our own ancestor, developed an internal molecular skeleton, the cytoskeleton.

The emergence of a cytoskeleton

The cytoskeleton is formed of two main classes of molecule, actin filaments and microtubules. They perform complementary functions: actin filaments resist pulling forces, and microtubules resist compression and shearing forces. These properties enable the cytoskeleton to maintain the form of the cell, in the absence of a rigid cell wall. But the cytoskeleton can also change the shape of the cell and move things around within it. Thus microtubules are used as rails in intracellular traffic of particles and vesicles. They also pull chromosomes apart in cell division, and are constituents of such motility organs as the cilia—complex whip-like structures whose beating drives cells through a fluid medium (sperm cells are driven by cilia). Actin filaments are active in cell division and in phagocytosis. To move things around, molecules must exert mechanochemical activity: they must convert chemical activity into mechanical motion. This requires that protein molecules should be able actively to change their shape: in effect, they must be able to stick out an 'arm', grab hold of something, and pull.

Where did this ability come from? It exists in a rudimentary form already in the eubacteria. When a bacterium divides, a furrow is formed in the cell membrane. This requires a mechanically active molecule: the gene sequences of some eukaryotic mechanochemical proteins show some resemblance to this bacterial molecule. Thus the 'fission-aiding' protein turned out to be preadapted to a cytoskeletal function.

In modern eukaryotic cells, there are several different kinds of membranes.

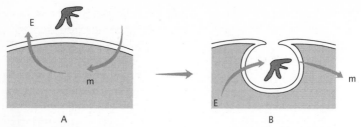

A   B

Figure 6.3 The origin of phagocytosis. In prokaryotes (A), solid food particles can be digested only by secreting digestive enzymes (E) into the surrounding medium, and absorbing the small molecules (m), resulting from digestion. Phagocytosis (B) could have originated simply by forming a vacuole, because the membrane would already be adapted to secrete enzymes into the vacuole, and to absorb the products of digestion. The existing mechanism, shown in Fig. 6.2, with a division of labour between food vacuoles and lysosomes, could have evolved later.

Our next topic is how these could have evolved. But the initial evolution of the food vacuole, budded off by cytosis from the cell membrane, presented no great difficulty. In bacteria, the digestive enzymes to be secreted were synthesized by ribosomes attached to the cell membrane. If, in the first eukaryotes, a food vacuole was formed accidentally by the engulfment of a food particle, aided by the rudimentary cytoskeleton (the furrow-forming bacterial molecules), then the digestion of food within the vacuole, and the takeup of nutrients, would be carried out by the original bacterial mechanisms (Fig. 6.3).

The endomembrane system

The eukaryotic cell contains many different kinds of membranes. We cannot describe them all here, but we will say enough to give a feel for what is involved. A convenient place to start is with a food vacuole. In bacteria, there are ribosomes attached to the outer cell membrane, synthesizing proteins that form part of that membrane—for example, proteins involved in the passage of nutrients and waste products through the membrane—and also digestive enzymes secreted into the surrounding medium. At the end of the last section, and in Fig. 6.3, we supposed that, when a food vacuole was formed in a primitive eukaryote, it resembled the cell membrane in being able to synthesize digestive enzymes and absorb nutrients. This is no longer true of existing eukaryotes. There are no ribosomes attached to the cell membrane. Consequently a food vacuole formed from the cell membrane cannot digest the food it contains. Before food can be digested, the vacuole must fuse with a lysosome, which does contain digestive enzymes, as was shown in Fig. 6.2.

If the cell membrane lacks ribosomes, and hence the ability to synthesize proteins, how does it grow? Essentially, it grows by a process which is the reverse of

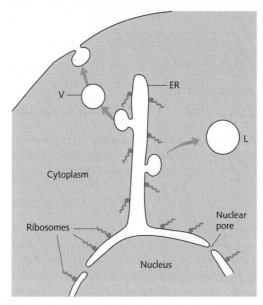

Figure 6.4 The endoplasmic reticulum (ER) and nuclear membrane. Note that the ER is continuous with the outer of the two membranes surrounding the nucleus: both have attached ribosomes and synthesize proteins. V is a vacuole budded off the ER, which will fuse with the cell membrane and contribute to its growth. L is a lysosome, also budded off the ER.

that by which a food vacuole is formed (process B rather than A in Fig. 6.1): a vesicle formed within the cell fuses with the outer membrane. These vesicles originate from a structure known as the endoplasmic reticulum, or ER (Fig. 6.4). This is a system of double membranes within the cell: it consists of two membranes pressed together, because it is formed by the flattening of spherical vesicles. Ribosomes are attached to the ER, so proteins can be synthesized. Vesicles of various kinds are budded off the ER, with the appropriate proteins embedded in them. Among these vesicles are those that fuse with the outer membrane and contribute to its growth, and also the lysosomes.

In eukaryotes, then, there is differentiation between the cell membrane, lacking ribosomes, and the ER, which has them, and which is therefore able to synthesize new cell membrane material. This may have been the first difference between membranes to evolve, but it has been followed by many others. The envelope surrounding the nucleus, like the ER, is formed from flattened vesicles, and so consists of two adpressed membranes. The inner one has no ribosomes (there is no protein synthesis inside the nucleus); the outer of the two membranes is continuous with the ER, as shown in Fig. 6.4.

With the formation of the nuclear membrane, there was a division of labour between the inside of the nucleus, where there is transcription, producing mRNA, but no protein synthesis, and the cytoplasm, where protein synthesis occurs. This requires pores in the nuclear membrane, through which mRNA can pass out to the cytoplasm, and those enzymes needed for DNA replication and transcription can pass from cytoplasm to nucleus.

The existence within the cell of different kinds of membrane, with different kinds of protein embedded in them, raises a problem. How do particular proteins find their way to the right places? For example, how do proteins needed in the nucleus find their way to the nuclear pores? The general answer is that such proteins have specific 'signal' peptide sequences with an affinity for recognition sites on the appropriate membrane. We will revisit this problem when discussing the evolution of organelles because this, too, requires that particular proteins finish up in the right places.

The origin of mitosis

When a cell divides, it is crucial that one copy of the genetic material is transmitted to each daughter cell. The mechanisms whereby this 'chromosome segregation' is achieved are very different in prokaryotes and eukaryotes. The prokaryotic mechanism is shown in Fig. 6.5, and the more familiar eukaryotic process of mitosis in Fig. 6.6. They are so different that it is hard to imagine a transition between them. But Tom Cavalier-Smith has suggested a plausible scenario, outlined in Fig. 6.6. The details are complicated, so we will first give an outline of the main points.

In prokaryotes, chromosome segregation depends crucially on the attachment of the origin and terminus of replication to the cell wall. In particular, it depends on the positioning of the new origin of replication ('O' on Fig. 6.5) opposite to the old one (O). It is impossible, therefore, for several origins of replication to exist on the same chromosome. The whole chromosome is therefore one replication unit, or replicon. This sets an upper limit to the rate of cell division. It takes about 40 minutes for a replisome to travel from origin to terminus, when conditions are optimal. If only one replisome was active at a

---

Figure 6.5 Bacterial cell division. (i) The cell before division, with the circular chromosome attached at the origin and the terminus to the cell wall–cell membrane complex. (ii) Replication proceeding in both directions: the new origin and the replication complexes (R) are shown exploded, but, as shown in (b₂), the new origin is bound to the replicases, and so is carried to the terminus. (iii) Replication is complete, and the new origin is attached to the cell wall. (iv) The terminus is re-attached to the cell wall midway between the two origins: this movement is brought about by coiling of the chromosomes. (v) The terminus divides, and the formation of a septum leads to cell division.

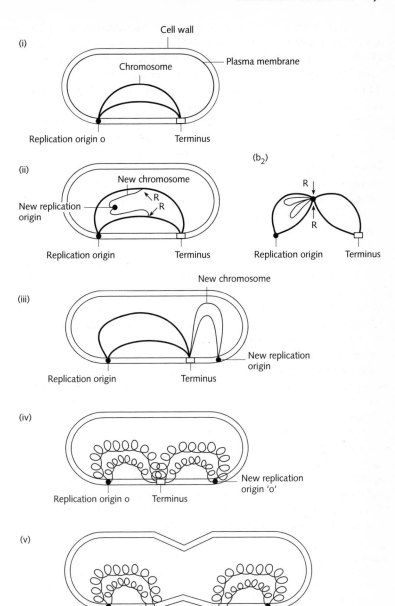

(i)

Cell wall

Chromosome

Plasma membrane

Replication origin o

Terminus

(ii)

New chromosome

New replication origin

R

R

Replication origin

Terminus

(b₂)

R

R

Replication origin

Terminus

New chromosome

(iii)

New replication origin

Replication origin

Terminus

(iv)

New replication origin 'o'

Replication origin o

Terminus

(v)

Replication origin

Terminus

New terminus

New replication origin

time, this would place an upper limit on the rate of reproduction at once every 40 minutes. Bacteria can do better than this—in fact, about twice as well—by having more than one replisome travelling along the chromosome at a time. But there is a limit to what can be achieved in this way, because of 'gene dosage'. Thus if two replisomes are active at the same time, there are four copies of genes close to O, but only one copy of genes close to the terminus. But gene dosage has important effects on metabolism, so a bacterium cannot permit too many replisomes to be active simultaneously. This places a limit on the rate of replication for a genome of a given size or, in evolution, a limit on the size of the genome that is compatible with rapid cell division.

The loss of a rigid cell wall forced the ancestral eukaryotes to evolve a new way of segregating their chromosomes. In this new mechanism, mitosis, the chromosomes are pulled apart by the microtubules to which they are attached. Microtubules are not present in the prokaryotes: they are a component of the cytoskeleton, which evolved to compensate for the loss of the cell wall. From one point of view, therefore, mitosis can be seen as something forced on the eukaryotes because the old, prokaryotic mechanism was no longer effective. Mitosis, however, also opened up new possibilities for the eukaryotes. Because the machinery ensuring that one chromosome passes to each daughter cell no longer depends, as it does in bacteria, on the attachment of the origin and terminus of replication to the cell wall, there is no longer any constraint on the number of origins of replication on a single chromosome. In eukaryotes, each chromosome has many replication origins. There is no limit, therefore, on the

---

Figure 6.6 The evolution of mitosis: (iii) shows the mechanism of mitosis in most existing eukaryotes. For simplicity, the figure shows only one chromosome, although varying numbers are in fact present. The chromosomes are moved apart by the mitotic spindle, a structure formed of microtubules. Each chromosome is attached to the spindle by a structure called the centromere. When the centromere has been replicated, the two copies are pulled to opposite poles of the spindle: these poles are called centrosomes. The movement is produced because two tubules lying side by side can actively slide past one another. How could this mechanism have evolved from the bacterial mechanism, shown in Figure 6.5? Diagrams (i) and (ii) show possible intermediate stages: (ii) is a form of division, called pleuromitosis, found in some existing protists; (i) is hypothetical. A possible scenario is as follows. In the hypothetical intermediate (i), the chromosome is still circular, and the origin and terminus are still attached to the cell wall, as in bacteria. The main innovation is that the origin of replication has divided, and new and old origins move apart while chromosome replication is still in progress. Pleuromitosis (ii) differs in several important respects. The chromosome is linear, not circular: the bacterial origin of replication has evolved into the centrosome, and the terminus into the centromere. The mechanism of division now depends on microtubules, and not on attachment to the cell wall. As in modern mitosis, the two centrosomes are driven apart by microtubules, and each centromere is pulled towards its own centrosome by microtubules. The evolution from pleuromitosis to typical mitosis is then easy to understand: the topological similarity is shown in (iv).

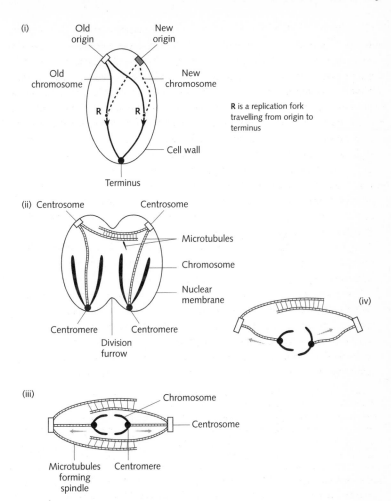

(i)

Old origin

New origin

Old chromosome

New chromosome

R is a replication fork travelling from origin to terminus

R   R

Cell wall

Terminus

(ii) Centrosome   Centrosome

Microtubules

Chromosome

Nuclear membrane

Centromere   Centromere

Division furrow

(iv)

(iii)

Chromosome

Centrosome

Microtubules forming spindle   Centromere

quantity of DNA per genome, or at least no limit set by the rate of replication. In eukaryotes, the DNA content can increase by several orders of magnitude over that present in bacteria: the human genome contains some 10^9 base pairs, compared with 10^6 in a coli bacterium.

This has interesting implications for the evolution of increasing complexity. The increase in the permitted size of the genome, brought about by the change in division mechanism, was a necessary, although not sufficient, condition for the evolution of complex multicellular organisms. But the change in division mechanism did not evolve *because* it permitted this future evolutionary

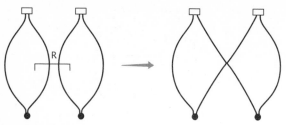

Figure 6.7 Crossing over between circular chromosomes. A recombination (R) between two circular chromosomes inhibits cell division.

potential: evolution does not have that kind of foresight. As we have suggested, the new division mechanism was forced on the first eukaryotes, but once it had evolved the evolution of increased complexity was possible.

But why the replacement of circular by rod-shaped chromosomes? The likely explanation is that, if recombination occurs between two circular chromosomes, this prevents their proper separation in cell division (Fig. 6.7). But why should recombination have occurred? Normally, we think of recombination as being important in the sexual process, forming new genetic types. But what is the point of recombination in mitosis? There is no point in having sex with oneself. The answer is probably that mitotic recombination was an unselected by-product of DNA repair. Sometimes, high-energy events damage both strands of a DNA molecule: when such damage is repaired, the damaged section is replaced by a new piece of DNA, copied from the undamaged homologous chromosome. The enzymes needed for such repair are present in bacteria as well as in eukaryotes. The snag is that, when such repair occurs, there is sometimes recombination between the two chromosomes: this would have had disastrous consequences if the chromosomes were circular, as explained above, but would not matter if chromosomes were rod-shaped. We will revisit this phenomenon when discussing the origin of sex.

There remains the question of how the nuclear envelope originated. A likely answer is illustrated in Fig. 6.8.

From symbionts to organelles

We now turn to the origin of mitochondria and chloroplasts, which, on the evidence of molecular sequence data, are believed to be descended from symbiotic purple bacteria and cyanobacteria, respectively. Endosymbioses are by no means rare. For example, animals of deep-sea vents have symbiotic bacteria that produce organic matter from carbon dioxide by using energy gained from the oxidation of hydrogen sulphide emanating from the vents. The evolution of

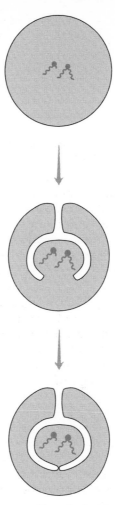

Figure 6.8 The nuclear envelope: a possible mode of origin, by invagination from the outer cell membrane. This scenario is consistent with the fact that the nuclear envelope is double, and that the outer membrane is continuous with the endoplasmic reticulum, as shown in Fig. 6.4.

such symbioses is discussed in Chapter 9. But the ancient organelles differ from these more recent examples of symbiosis in one important respect: most of the genes once present in mitochondria and chloroplasts have been transferred to the cell nucleus. We discuss two questions. First, what did eukaryotes gain through symbiosis with the ancestors of mitochondria and chloroplasts?

Second, what were the causes and consequences of the transfer of genes from organelle to nucleus?

The role of organelles

What do eukaryotic cells get from their organelles? To answer this question, one must understand something about chemical energy—where it comes from and what it is used for. In modern society, our energy comes in the main from fossil fuels, coal (carbon) and oil (a compound of carbon and hydrogen). When burnt (oxidized) to carbon dioxide (CO_2) and water (H_2O), energy is released, which can be used for driving machinery: burning coal drives steam engines, and burning petrol drives motor cars. Much the same is true of ourselves. We gain energy by burning sugars and other carbohydrates. In the first instance, we use this energy to synthesize molecules of ATP. It is helpful to imagine an ATP molecule as being like a jack-in-the-box. It can be carried round the cell; whenever energy is needed, the jack is released and the energy that becomes available as the spring extends is used to drive some other process. Energy, obtained by oxidizing sugars, is then needed to push the jack back into the box again. Roughly, this is what is meant by saying that ATP is the cell's source of 'free energy'. This energy can be used to drive many processes in the cell that require energy—for example, synthetic processes, cell movement, pumping molecules across membranes, and so on.

Early in the evolution of life, organic compounds, including sugars, were synthesized abiotically (see p. 31). There was no free oxygen then, however, so energy could not be obtained by oxidizing these organic compounds. But it is possible to get some energy from sugars without oxygen, by a process called fermentation, whereby sugars are converted to alcohol: in wine-making, for example, the sugar in grapes is converted to alcohol by yeast cells. The process does yield some energy, but only about one-tenth of what could be obtained by oxidizing the same quantity of sugar.

Today, life is not dependent on abiotically produced compounds. Instead, green plants trap the energy of sunlight, and use it to split water into hydrogen and oxygen. The oxygen is released into the atmosphere and the energy available in the free hydrogen is used to synthesize sugars.

Both these processes—the trapping of sunlight to synthesize sugars and the oxidation of sugars to synthesize ATP—were already present in the prokaryotic world. In eukaryotes, they are carried out in organelles. ATP synthesis is carried out in the mitochondria, which are known from molecular evidence to be descended from the purple non-sulphur bacteria, and photosynthesis is carried out in the chloroplasts, descended from the cyanobacteria. There is a puzzle here. Both bacterial ancestors, purple bacteria and cyanobacteria, can carry out both functions. So why were two symbiotic events needed? The answer lies in comparative efficiencies. The purple bacteria have an inefficient version of photo-synthesis, and can live only in environments containing particular sulphur compounds. The cyanobacteria, although efficient at photosynthesis, are less so at respiration.

How do these organelles benefit the cell? To be any use, their products must pass out of the organelle into the cell cytoplasm. This would not have happened in their free-living ancestors: it would be wasteful for purple bacteria to lose ATP, or for cyanobacteria to lose organic compounds, to the outside medium. Today, the transport of energy and organic compounds depends on the pres-ence in the organelle membrane of specific transport proteins, or 'taps'. This raises the problem, which is not yet fully solved, of how the first endosymbionts were useful to the ancestral eukaryotes, before the evolution of special transport proteins to make their products available.

The evolution from symbiont to organelle involved the transfer of many genes to the cell nucleus, followed by the loss of those genes from the genome of the organelle. Today, mitochondrial genomes contain about a dozen genes, and plastids from one to two hundred. Obviously, thousands of genes could not have been lost from the symbionts before functional transfer to the nucleus was complete; otherwise, the symbiont, and consequently the whole cell, would have become inefficient or inviable. It was not sufficient for a gene to be trans-ferred to the nucleus for it to be lost from the symbiont. Indeed, such transfers may not have been particularly difficult: there is evidence that DNA moves be-tween compartments in the cell (nucleus, plastids, mitochondria) rather easily. But it was also necessary that the proteins produced by those genes should find their way back to the appropriate organelle. This is achieved by a special string of amino acids at the end of the protein, forming a 'transit signal' that is recog-nized by the membranes of the organelle, and facilitates transfer of the proteins through the membrane.

Each protein thus imported carries the signal sequence at the end that first emerges from the ribosome during synthesis. A hint as to the origin of these sequences comes from the observation that they are not the same in different proteins, although they are sufficiently similar to be recognized by the same receptor. This suggests that each protein evolved its signal sequence independ-ently. Of course, each organelle recognizes and imports only its own proteins.

In Chapter 1, we distinguished between 'unlimited' heredity, as possessed by DNA, and 'limited' heredity, whereby a replicator can exist in one of a relatively small number of states, each of which reproduces its kind. It is a fascinating fact that cell membranes have limited heredity in this sense. For example, mito-chondrial membranes contain proteins that catalyse the import from the cyto-plasm of all the appropriate proteins, including the receptor and transport proteins themselves. The same is true of other membranes. This has prompted

Tom Cavalier-Smith to introduce the concept of membrane heredity: mitochondrial membranes facilitate the synthesis of further mitochondrial membrane, plastid membranes of further plastid membrane, and so on. This
hereditary potential is independent of both the nuclear and organellar genes,
as one can appreciate from the following thought experiment. Suppose that, in
a mitochondrion, one replaced the mitochondrial receptor channel by the
analogous plastid molecules. This operation would cause plastid proteins to be
incorporated into the mitochondrion. Although neither the organellar nor the
nuclear genome was altered, the mitochondrion would undergo hereditary
deterioration. Membrane heredity is limited, but is of unlimited importance to
eukaryotic cell function.

We end this account of organelle evolution by describing the situation in a
strange group of algae, the Chromista, to illustrate how complex the results can
be. Today, these algae have genetic material derived from four different ancestors.
This is most easily explained by describing their evolutionary history. First, it
seems that a green algal cell lost its chloroplasts, and hence its ability to photosynthesize. Later, its descendants re-acquired this ability by engulfing a new
symbiont: not, this time, a prokaryotic cyanobacterium but a eukaryotic red
alga. This gave rise to cells that, today, contain four distinct genomes. The main
nuclear genome is that of the host cell, originally a green alga, which also contained mitochondria, descended from purple bacteria. Within these cells there
still survive some chromosomes derived from the engulfed red alga, and, finally,
the chromosomes of the chloroplast, provided by the red alga, but originally
descended from a free-living cyanobacterium.

The changing picture of eukaryotic evolution

It may be that the picture we have drawn of the earliest eukaryotes and the
acquisition of mitochondria will have to be changed in the light of recent discoveries, published when this book was almost completed. We present these
ideas in this section, because they are still tentative. Fortunately, even if they
prove to be correct, most of the points made above remain unchanged.

As we explained, existing eukaryotes can be divided into two groups: typical
eukaryotes containing mitochondria, and a second group, the archaezoans,
lacking them. In our account, we assumed that at least some archaezoans are
descended from ancestors that never had them: they represent the primitive
eukaryotic state of a cell with a nucleus, with a cytoskeleton but no rigid cell
wall, feeding by phagocytosis, and lacking mitochondria or other organelles.
Recent research suggests that all existing archaezoans may once have possessed
mitochondria, and have since lost them because they live in environments lacking oxygen—for example, as parasites. The reason for thinking that this is so is

that their nuclei contain some genes whose only plausible origin is from mitochondrion-like symbionts.

By itself, this observation need not change the scenario we described. It may be that the first eukaryotes did indeed lack organelles, but have left no living descendants. But there is an alternative scenario. This is that, from the very beginning, there was a close association between two very different kinds of bacterial cell, one an archaebacterium and one a eubacterium, but the metabolic significance of the association was different from the one we described above.

Miklós Müller from the Rockefeller University has suggested a novel approach (Fig. 6.9). First, he emphasizes—following others—that the genes of the eukaryotic nucleus seem to have a double origin. Genes concerned with DNA replication and protein synthesis resemble those of archaebacteria, whereas many

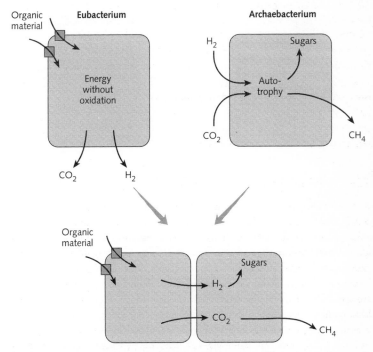

Figure 6.9 The original association between a eubacterium (left) and a methanogenic bacterium (right), as hypothesized by Miklós Müller. The diagram shows the chemical co-operation that would have occurred in an environment without oxygen; but the eubacterium would also be able to respire in aerobic conditions. The open squares indicate a protein transporting organic compounds across the cell membrane.

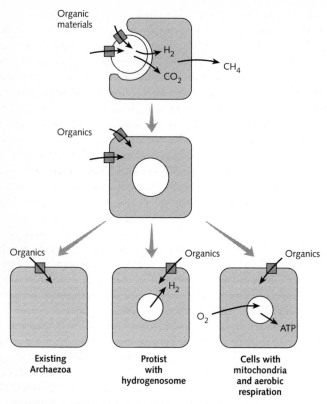

Figure 6.10 Further evolution of the original association between two bacteria. As in Fig. 6.9, the open squares indicate a membrane transport protein.

metabolic genes, including those encoding the enzymes for fermentation, resemble those of eubacteria. Müller suggests that one of the two symbionts was a 'methanogenic' archaebacterium. A methanogen does not require oxygen, and obtains energy by taking in hydrogen and carbon dioxide, and releasing methane (CH_4)—hence the name methanogen. The other symbiont, he suggests, was a eubacterium, able to obtain energy in the absence of oxygen by fermenting sugars (this is the crucial difference from the earlier scenario, in which the symbiont contributed the ability to get energy by oxidizing sugars). What did the eubacterium do for its archaebacterial partner? The idea, illustrated in Fig. 6.10, is that it produced carbon dioxide and hydrogen as a by-product of its metabolism: these are precisely the compounds needed by the methanogen for the synthesis of organic compounds.

This idea gets round one of the difficulties we mentioned in our earlier account, *The major transitions in evolution*. We pointed out that, before a symbiont producing ATP would be of any use to its host, some means of transporting ATP across the symbiont membrane would be needed. No such difficulty arises with Müller's idea. But, in removing one difficulty, the new suggestion raises another. If, as we proposed earlier, the original host cell first evolved the habit of swallowing things by phagocytosis, this would explain how the symbiont got inside the host. But a methanogen has no need to swallow things: hydrogen and carbon dioxide will diffuse across the cell membrane. So we have to suppose (Fig. 6.10) that the symbiosis started as a simple contact between the two cells, followed by a gradual change in the shape of the archaebacterial host, leading to the engulfment of the symbiont: this would make the transfer of materials more efficient (such changes in the shape of the cell envelope do occur in some spore-producing bacteria).

In different evolutionary lineages, the eubacterial symbiont had various fates (Fig. 6.10). In some cases it was completely lost, giving rise to the existing archaebacteria. In others, it evolved into an organelle called a hydrogenosome, which lacks DNA and produces hydrogen: there are existing protists that lack mitochondria but possess hydrogenosomes. Finally, in one lineage it evolved the capacity to utilize oxygen: that is, it evolved into a mitochondrion.

These new ideas explain recent observations on the dual ancestry of eukaryotic genes, and get round some of the difficulties of our earlier account. But they do so at the expense of complicating the origin of the cytoskeleton and phagocytosis. We do not know which scenario is nearer to the truth. The British geneticist Laurence Hurst once said that, if he were given a wish, he would ask for the true phylogeny of all known organisms. Unfortunately, we do not own the 'Book of Phylogeny' but must infer it by fallible methods. Faced with the problem of eukaryotic origins, we cannot help but sympathize with Hurst's view.

Conclusions

Although we have written of the origin of the eukaryotes as one of the 'major transitions', it was in fact a series of events: the loss of the rigid cell wall, and the acquisition of a new way of feeding on solid particles; the origin of an internal cytoskeleton, and of new methods of cell locomotion; the appearance of a new system of internal cell membranes, including the nuclear membrane; the spatial separation of transcription and translation; the evolution of rod-shaped chromosomes with multiple origins of replication, removing the limitation on genome size; and, finally, the origin of cell organelles, in particular the mitochondrion and, in algae and plants, the plastid. Of these events, at least the last two qualify as major transitions, in the sense of being changes in the way genetic information is stored and transmitted.

The fascinating thing about this story is the way in which many apparently unconnected changes, setting the scene for all subsequent evolution, were in a sense forced on the cell by the loss of the cell wall, an event that might have seemed at the time both trivial and regressive.

CHAPTER 7

THE ORIGIN OF SEX

In animals and plants, and in eukaryotes generally, the essence of the sexual process is that a new individual starts from a single cell, the 'zygote', formed by the fusion of two sex cells, or 'gametes'. Typically, gametes have only one set of chromosomes—that is, they are 'haploid'—and consequently the zygote contains two sets of chromosomes—it is 'diploid'. Hence a new individual contains genetic information from two parents, and, in the longer term, each individual has received genes from many ancestors, and may contribute genes to many descendants. This has led to the concept of a 'gene pool', consisting of genes that may be in separate individuals today, but whose ancestors may have been together in the same individual in the past, and whose descendants may be together again in the future. The species, or set of potentially interbreeding individuals, thus constitutes an evolving unit, with a common gene pool.

The first point to make is that, although biologists often speak of 'sexual reproduction', the sexual process is in fact the precise opposite of reproduction. In reproduction, one cell divides into two: in sex, two cells fuse to form one. Sex is not even necessary for continued reproduction. Many single-celled organisms, and some animals and plants, reproduce indefinitely without sex. The production of eggs that develop without fertilization is called parthenogenesis, or virgin birth. Many insect species consist only of parthenogenetically reproducing females. Among reptiles, there are parthenogenetic species consisting entirely of females producing daughters genetically identical to themselves. One of the American whiptail lizards, *Cnemidophorus uniparens*, is such a species: it is thought to be of relatively recent origin, perhaps thousands rather than millions of years old, and to be descended from a female that was probably a hybrid between two sexual species. Parthenogenesis is even commoner in plants: for example, most dandelions, blackberries, and ladies' mantles reproduce without sex. It is curious that mammals never reproduce parthenogenetically, and that there are no parthenogenetic species of birds, although parthenogenesis is not unknown in domestic varieties. Thus, whatever may be the explanation of sex, it cannot be said that without it continued reproduction is impossible.

The problem

We are accustomed to associating sex with sexual differentiation—that is, with a difference between males and females. In animals and higher plants, this is justified. In animals, males produce small motile gametes (sperm), and females large non-motile ones (eggs). In hermaphrodite species—for example, many snails and flatworms—the same individual produces both eggs and sperm. But differentiation of the gametes, and hence of males and females, is not a universal feature of sexual reproduction. Most single-celled eukaryotes produce gametes of only one size—they are 'isogamous', in contrast to the anisogamous animals and higher plants.

Our problem is to explain why sex arose, and why it is today so widespread. If it is not necessary, why do it? The problem is made harder by what has been called the 'twofold cost of sex'. To understand this cost, imagine a typical sexual species of lizard. A female can lay, perhaps, a hundred eggs during her lifetime, but on average, because the number of lizards remains roughly constant, only two of them will survive to breed, one a male and one a female. Thus, on average, each female will produce one daughter. Now imagine a mutant gene causing a female to be parthenogenetic, producing daughters genetically identical to herself. She, too, will, on average, lay a hundred eggs, of which two will survive. But both these will be parthenogenetic females. Initially, and barring accidents, the number of parthenogenetic females in the population will double in every generation. Rather quickly, parthenogens will replace sexuals. Thus there is a twofold advantage associated with parthenogenesis, or, equivalently, a twofold cost of sex. Of course, in this argument we have ignored any counterbalancing advantages that may be conferred by sex. Part of the problem of the evolution of sex is to identify those counterbalancing advantages.

In the next section we will discuss what those advantages might be. But, first, there is an important point to make about the twofold cost. We assumed above that a parthenogenetic female can produce as many offspring as a sexual one. The assumption is reasonable for lizards, in which there is no parental care, but would be less plausible in passerine birds in which both parents care for the young. The point is that males contribute no nutrients to the fertilized egg, and so, in the absence of male parental care, they are expendable. The situation is very different in isogamous species, in which the two gametes contribute equally, in nutrients as well as genes, to the new individual. In an isogamous species, a parthenogen would have to contribute twice as much to each new egg, if that egg was to have as good a start in life as a sexually produced one. Thus in isogamous species there is no necessary twofold cost of sex.

Because the first sexual eukaryotes were certainly isogamous, it follows that the twofold cost is a problem only if we are concerned to explain the maintenance of sex in later, anisogamous organisms, but not when discussing the origin of sex. All the same, there must be some costs associated with sex, even in isogamous organisms. Apart from the necessity of a gamete finding a partner with which to fuse, growth and reproduction are interrupted by the complex process of meiosis whereby gametes with half the number of chromosomes are produced.

To ensure the proper distribution of chromosomes, the production of gametes is a complicated process, as anyone familiar with the accounts of meiosis in biology textbooks will be aware. Because of these complications, and the obvious disadvantages associated with them, it is not surprising that the origin and maintenance of sex continue to be a matter of controversy among biologists.

The advantages of sex

We now seek selective advantages of sex that might counterbalance the costs associated with the interruption of growth to produce gametes, the finding of a partner with which to fuse, and, in higher organisms, the twofold cost to females of producing males. Many suggestions have been made. These need not be mutually exclusive: after all, sex might confer more than one benefit. We will not list them all, but it would give a misleading picture of the present state of understanding if we were to concentrate on a single answer. If you find the story confusing, welcome to the club.

A major difficulty concerns the level at which selection is acting. Does selection act between individuals or between populations? We described this debate on pp. 19–21: the outcome was to persuade most biologists that, usually, between-population selection would be overwhelmed by between-individual selection, and that there was in any case little need to invoke it. But there remained the problem of sex, which had traditionally been explained by the advantages it confers on populations. More recently, there have been strenuous efforts to explain the evolution of sex without invoking group-level selection. In fact, things are more complicated than we had thought. Selection can act at any one of three levels:

1. Sex may benefit some populations at the expense of others, although it is of no advantage to a sexual individual relative to an asexual one within a population. For example, it may speed up evolution or reduce the load of deleterious mutations. Such advantages are long-term ones.

2. Sex may benefit individuals. For example, a sexual female produces offspring that are not all alike: if competition is intense, and only some genotypes have a chance of survival, this may pay. Such an advantage would be effective in the short term, in one or two generations.

3. Sex may benefit some genes at the expense of others in the same individual: we will explain this idea below.

Sex benefits populations

There are two ways in which this may be the case. The first is that a sexual population can evolve more rapidly to meet a changing environment. The reason is illustrated in Fig. 7.1. Suppose that two mutations, $a{\to}A$ and $b{\to}B$, are both selectively advantageous. Typically, the two mutations will occur in different individuals. Barring accidents, both will increase in frequency. In a sexual population, recombination can bring the two mutations together in a single individual: soon, the whole population will be AB. In an asexual population, this cannot happen. An AB individual can arise only when a $b{\to}B$ mutation occurs in an individual that is already A, or vice versa. Calculations show that the effect on evolution rate can be substantial.

But this explanation requires that the environment should continuously be changing, forcing species to evolve to meet the challenge. Is it really true that the environment changes that rapidly? One idea is that the 'environment' of each species consists of other species: its competitors, predators, and parasites. When any one species changes, this is experienced by other species as a change in their environment, inducing them to change, and so on. The result is an arms race.

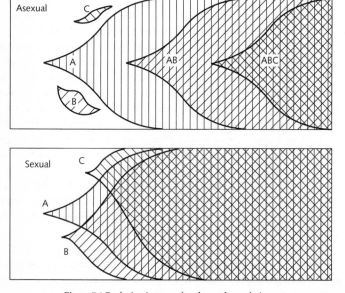

Figure 7.1 Evolution in asexual and sexual populations.

A second possible advantage is that sex may reduce the load of deleterious mutations in the population. Suppose that two individuals suffer from different harmful mutations, m_1 and m_2 . If they mate, they can produce a normal offspring, with no mutations, by recombination: without sex, they can do so only in the unlikely event of a back mutation. We will call this the 'engine-and-gearbox' theory. You can buy two clapped-out motor cars, one with a broken gearbox and one with a broken engine, and make one functional car. The snag is that the child of a mating between individuals with different mutations might have both mutations, rather than neither. As Bernard Shaw said to the actress who suggested they have a child, which would have her beauty and his brains, 'Yes, madam, but what if it had my beauty and your brains?'.

Despite this snag, there are circumstances in which a sexual population will carry a smaller load of deleterious mutations than an asexual one. If you are happy to take this on trust, you should skip the rest of this paragraph, and the next. There are two situations in which the statement is true. First, consider a finite population, subject to continuous, slightly harmful mutations. Individuals can be classified as having 0, 1, 2, etc. harmful mutations. Those with no harmful mutations belong to the 'optimal' class. In a finite population, particularly if it is small, there is a chance, every generation, that no individuals belong to this optimal class. If there is no sex, the class cannot be reconstituted: a new optimal class, with one harmful mutation, now exists. It, too, may be lost by chance, and so on. The process is known as 'Muller's ratchet', after Hermann J. Muller, the American geneticist who first suggested it. Sex, of course, can arrest the ratchet, and prevent continuing deterioration.

Sex may help even in an infinite population, but only if deleterious mutations act 'synergistically'; that is, if it is OK to have m_1 or m_2, but not both (or, more generally, if mutations have worse effects in combination than would be expected from the sum of their effects on their own). It is difficult to see why this should be so without some mathematics. No one, however, doubts the conclusion: if deleterious mutations act synergistically, sex and recombination reduce the load of deleterious mutations. But do deleterious mutations act synergistically? There are reasons, both theoretical and empirical, to think that they do, but more evidence would be welcome.

There is no difficulty, then, in thinking of reasons why sex and recombination might benefit populations. The difficulty arises because the advantages would be long-term, and therefore cannot explain the *origins* of sex, even if they may help to explain its maintenance. If parthenogenetic females have a short-term advantage, they are likely to replace sexual ones, resulting in an asexual population, even if in the long-term this is harmful to the population. The problem is particularly severe in higher organisms, with a twofold cost of sex. This is the classic difficulty with 'group selection' explanations: short-term advantages to the individual are likely to outweigh long-term advantages to the population. Nevertheless, there is a reason for thinking that group selection, favouring sexual populations, may have been important in the maintenance of sex. This comes from the taxonomic distribution of parthenogens. There are many parthenogenetic varieties and species, some parthenogenetic genera, but almost no higher taxonomic groups (families, orders) that are wholly parthenogenetic. This is exactly what we would expect if new parthenogenetic varieties arise from time to time, but are eliminated by between-group selection before they can produce a larger taxon.

It looks, therefore, as if group selection has been important in maintaining sex in higher organisms. But some reservations are needed. First, there are a few higher taxa that are wholly parthenogenetic. The most famous are the bdelloid rotifers, a whole order in which males have never been observed. It is important to find out what, if anything, is peculiar about these exceptional cases. Second, the group-selection explanation works only if the origin of a new parthenogenetic variety is a rare event. This is probably true. As we mentioned on p. 25, mammals are never parthenogenetic. The reason is essentially trivial. There is a curious phenomenon called 'gene imprinting'. In mammals, some genes 'remember' whether they were inherited from the father or mother. In some tissues only the father's gene is active, and in others only the mother's gene. Because the genes are essential, every child must have a father and a mother: this rules out parthenogenesis. The point of this example is that a secondary adaptation— imprinting of genes—has been built onto the sexual process, and makes reversion to parthenogenesis difficult or impossible. We could have given other examples of such 'sexual hang-ups'.

A third reservation is that in some species the same individual produces offspring both sexually and asexually. An example is the pussytoes, *Antennaria*, of the family Compositae, in which the same flower head may produce asexual seeds as well as ovules that require fertilization by a pollen grain. In such cases there must be some short-term advantage in retaining sex: they are well worth further study.

A final reservation is crucial in a book concerned with origins. Group selection may be relevant to the maintenance of sex, once it has arisen, but cannot be relevant to its origin, which requires short-term advantages.

The benefit of sex to individuals

Why should an individual that reproduces sexually leave more surviving offspring? We have already mentioned one possible answer. Sexually produced offspring are all different, whereas parthenogenetically produced offspring are usually identical genetically. As the American George Williams pointed out, a parthenogenetic female is like a man who buys 100 tickets in a raffle, and finds that they all have the same number. It would be better, like a sexual female, to buy only 50 tickets, all with different numbers. Models of this process, which we will call the 'raffle-ticket' model, can be made to favour sex, but they do require that selection be very intense, and very unpredictable. We think that an explanation along these lines will work only in rather exceptional circumstances.

Sex may be selected at an individual level for other reasons. In the last section, we suggested that sex could be advantageous to a population in two ways. A sexual population can evolve faster, and deleterious mutations may accumulate in an asexual population. Analogous advantages may accrue to individuals. In a rapidly changing environment, sexually produced offspring are more likely to have characteristics adapting them to the new circumstances: this is particularly likely if the challenge comes from rapidly evolving parasites. Sexually produced individuals may also have a lower load of deleterious mutations. There is still debate about whether such individual-level advantages are sufficient to maintain sex, particularly in face of the twofold cost of producing males.

There is another possible explanation of sex, of quite a different kind, that is popular with molecular biologists. Although we think it is based on a misunderstanding, it points to a process that may have been important in the origin of sex, so it is worth explaining. The idea is that sex exists to make possible DNA repair. First, we must distinguish between 'mutation' and 'damage' to DNA. A mutation is a change of one DNA molecule to another DNA molecule with a different base sequence: this is *not* the kind of change we are now considering. 'Damage' is a change of a DNA molecule into something that is not DNA at all. If it is not repaired, such damage may be fatal: the damaged molecule cannot be replicated so that all information is lost. The problem is how to repair it. Enzymes can recognize the damaged piece and excise it, but how can they replace it by new DNA carrying the original message—that is, with the original base sequence? If only one strand of the double helix is damaged, it can be replaced using the information on the undamaged strand. But what if both strands are damaged? Repair is then possible only by copying an undamaged DNA molecule with the same message. If the cell contains two copies of its DNA, then double-strand damage can be repaired by copying the undamaged molecule.

Such double-strand repair does occur. It is found not only in eukaryotes but also in bacteria. How can this be, if bacteria have only a single chromosome? The answer is that, when we say that bacteria have only one chromosome, we mean that they have only one *kind* of chromosome, with only one genetic message. But, most of the time, a bacterial cell contains two identical chromosomes: if one is damaged, it can be repaired using information from the other. The point is that double-strand repair requires diploidy—two copies of the genetic message—but not sex. Double-strand repair is ancient and important. We shall invoke it in our scenario for the origin of sex in the next section. But we see it as an explanation for diploidy, not sex.

Sex and selfish genes

Most genes replicate only when the cell replicates. There are, however, transposable elements, or 'transposons', which jump, or transpose, to a new site in the genome, while leaving the original copy behind: after transposition, there are two copies, where there was only one before. Consider such a transposon in an asexual organism. By transposing, it can increase the number of copies of itself in the organism but not the number of organisms in which it is present. Its long-term survival depends entirely on whether its host organism leaves descendants. But suppose such a transposon could cause its host cell to fuse with another cell. It could then transpose to the chromosomes derived from the other cell. If the cell then divided, the transposon would be present in both daughter cells, although it entered from only one parent. In other words, by causing cell fusion, the transposon would give itself an additional means of increase. Such an element could spread through a population, even if, as would be expected, it reduced somewhat the fitness of its host. What has all this to do with sex? Donald Hickey and Michael Rose suggested in 1988 that sexual fusion originated, not because of any benefit to the cells that fused, but because it benefited a transposon that was present in only one of the two cells that fused, but which was transmitted to all their joint descendants.

The idea is attractive for two reasons. It provides an immediate selective advantage for sexual fusion: we need not appeal to the long-term advantage to the population. Further, there is a precedent for such genetic elements in bacteria. Plasmids are accessory genetic elements, usually carrying only a few genes, present in most bacteria. They cannot cause their host cells to fuse, fusion being difficult for a cell with a rigid cell wall, but some plasmids do cause the cell they are in to attach to another bacterium. When this happens, a copy of the plasmid DNA is passed to the other cell. Occasionally, DNA of the bacterial chromosome can also be passed to the recipient cell. So the scenario proposed by Hickey and Rose is not too far-fetched. If they are right, it leads to a somewhat paradoxical conclusion. Sexual fusion originated because it benefited a transposon, at the expense of the host cell. In the long-term, sex survives because it benefits the population, although, again, the short-term interests of the individual would be better served by parthenogenesis.

In the next section, we suggest a way in which sexual fusion could have originated. Our proposal does not depend on selfish transposable elements, but, as we explain, such elements could have played a part.

The various advantages for sex that we have proposed in this section are summarized in Table 7.1.

Table 7.1 Theories of the evolution of sex

---

1. Selection favours sexual populations
    (a) Sexual populations can evolve more rapidly
    (b) Asexual populations accumulate deleterious mutations

2. Selection favours individuals that reproduce sexually
    (a) It pays an individual to produce a variable progeny (the lottery model)
    (b) Sex makes repair of damaged DNA easier
    (c) Even within a population, selection may favour sexual individuals for the reasons
        under (1) above: in a changing environment, their offspring may be better adapted to
        the new circumstances, or, in an unchanging environment, their offspring may have
        fewer harmful mutations

3. Selection favours genes that cause individuals to undergo sexual fusion (or to produce
    gametes that fuse) because then a gene present in one of the fusing cells can transfer to
    the other.

A theory for the origin of sex

Figure 7.2 shows a possible scenario for the origin of sex. The first stage shows a single-celled organism, with a haploid–diploid life cycle. For part of the time the cells grow as haploids, with a single set of chromosomes, and at other times as diploids. Thus the population cycles between haploid and diploid states. In this respect it resembles some modern sexual organisms—for example, some seaweeds—but the means of changing from haploid to diploid, and from diploid to haploid, are different. The haploid becomes a diploid not by fusion with another haploid but simply by replicating each chromosome without dividing into two cells. This simple process, known as 'endomitosis', occurs in some cells today.

The diploid is converted into a haploid by a simplified version of meiosis. Because it is the origin of meiosis, which, it seems to us, is the hard thing to explain, we digress for a moment to discuss the problem. Before meiosis, each cell contains two copies of each kind of chromosome—two As, two Bs, and so on: after meiosis, each cell contains one of each kind. How is this to be achieved? To understand the answer, imagine that you are a schoolteacher with a class of children, of whom two are called Anne, two Bill, two Charles, and so on through the alphabet. To organize a game, you wish to divide the class into two teams, each containing one Anne, one Bill, one Charles, and so on. You could append to the wall a list of the two teams, with surnames, and tell the children to read the list, and assemble accordingly. But there is a much easier way. Tell each child to find the partner with the same name, and then to separate, one member of each pair going to one assembly point, and one member to the other. This is exactly what happens in meiosis. Each chromosome pairs with its similar partner, or 'homologue', and the members of the pair are then pulled

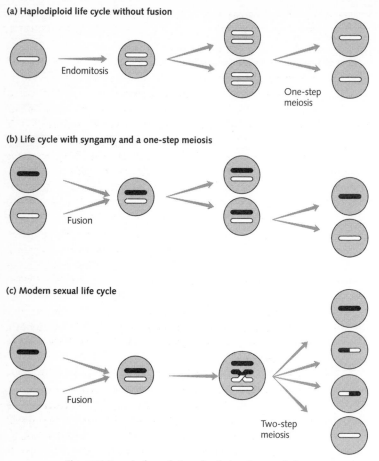

(a) Haplodiploid life cycle without fusion

Endomitosis

One-step meiosis

(b) Life cycle with syngamy and a one-step meiosis

Fusion

(c) Modern sexual life cycle

Fusion

Two-step meiosis

Figure 7.2 Stages in the evolution of meiosis and gamete fusion.

apart, one to each pole of the spindle. Effective meiosis depends on pairing between homologues. Interestingly, hybrids between species are often sterile because the chromosomes of the two parents are too different to pair.

In the first stage, illustrated in Fig. 7.2, we imagine a 'single-step meiosis', in which there is no initial doubling of the chromosomes, and no crossing over: these are complications that we discuss below. All that happens is that each chromosome pairs with its homologue, the pairs separate, and the cell divides. There would be no point in crossing over, because the two homologues are identical, or almost so, having arisen by endomitosis only a few cell generations ago. Such a one-step meiosis is not pure hypothesis. It has been described as occurring today in some single-celled protists. We do not know whether cross-ing over occurs in the meiosis of these protists: genetic studies are needed to find out.

The haploid–diploid cycle that we propose as an ancestral state is therefore quite plausible. Both endomitosis and a one-step meiosis are found in existing organisms. But what selective advantages could such a cycle confer? Presum-ably, it was an adaptation to a changing environment, perhaps to the annual cycle. There are good reasons why, in some environments, it would pay to be a diploid. As we explained earlier, damaged DNA can be repaired only if there is an undamaged molecule to copy: that is, it requires diploidy. DNA damage is more frequent if the oxygen concentration is high, so diploidy could be ex-plained as an adaptation to periods of high oxygen. But what were the corre-sponding advantages of haploidy? Haploid cells are smaller and, therefore, have a higher ratio of surface area to volume. For this reason, haploid yeast cells grow faster than diploids at low nutrient concentrations. But there is a snag to this explanation. Yeast cells feed by the diffusion of molecules through their surface. But we argued in the last chapter that the first eukaryotes fed by engulfing par-ticles. So we are faced with a problem. Perhaps the best hope lies in a study of present-day protists with a haploid–diploid cycle.

In a sense, the first stage that we hypothesize has nothing to do with sex, which requires that DNA from different ancestors be brought together in a single descendant. The virtue of suggesting such a stage is that it helps to explain the origin of a simple form of meiosis.

In the second stage in Fig. 7.2, endomitosis has been replaced by cell fusion as a means of restoring diploidy. This is a genuine sexual cycle. We can suggest two reasons for the change. The obvious one is that fusion would cover up the effects of recessive deleterious mutations. Two different haploid cells would carry different mutations: if the mutations were recessive, a diploid formed by the fusion of two such cells would not suffer the ill effects of either mutation. This is the mechanism that underlies the phenomenon of 'hybrid vigour', whereby hybrids between inbred lines are more vigorous than their parents. The effect would have been less striking than it is when crossing diploid inbred lines today, because seriously deleterious mutations would be eliminated by selection from haploid cell lines. But the advantages of fusion over endomitosis could still be substantial. A second possible reason for cell fusion was men-tioned on p. 86: a selfish transposon could have caused cell fusion as a means of ensuring its own propagation.

Once diploidy is restored by cell fusion rather than endomitosis, the two homologous chromosomes, although similar, will be slightly different, so that crossing over will produce new genotypes. In the third stage in Fig. 7.2, the simple one-step meiosis without crossing over has been replaced by a two-step meiosis with recombination. This raises two questions. What are the advantages of recombination? Why, in modern meiosis, is there an initial doubling of every chromosome, requiring two cell divisions to restore haploidy?

The advantages of recombination are now rather well understood. On pp. 81–85, we explained that a sexual population has two potential advantages over an asexual one. It can evolve faster, and may have a lower load of deleterious mutations. Suppose now that we have a sexual population, with cell fusion and meiosis, and suppose also that there are genes that alter the rate of recombination. The latter assumption is plausible: every sexual population that has been carefully studied has been found to have genetic variability in the rate of recombination. It turns out that the same circumstances that favour sexual populations over asexual ones will favour genes increasing the rate of recombination within a sexual population. If the environment is changing rapidly, or if deleterious mutations act synergistically, selection will favour increased rates of recombination.

The second question is harder to answer. Why a two-step meiosis? Every elementary textbook describes the curious process whereby, first, every chromosome replicates, so that there are four copies of each kind of chromosome: then crossing over occurs between homologous strands, so that genes inherited from different parents can be transmitted to the same offspring: and then two cell divisions produce haploid gametes. But the same texts rarely comment on how odd this process actually is. The primary function of meiosis is to halve the chromosome number: why then start by doubling it? It is not obvious that recombination requires a four-strand stage. As mentioned above, we do not know whether crossing over occurs in the one-step meiosis of archaezoans, but it is hard to see why it should not. We know of two suggested explanations of the two-step nature of meiosis. They are too complex to explain here, but are described briefly in Chapter 9 of our book *The major transitions in evolution*.

One final point about meiosis is important. The process of recombination is a complex one, requiring that two strands of DNA be lined up precisely, cut at precisely the same point, and rejoined with a change of partners. The process requires enzymes. How could such a complex system evolve, because it would be worse than useless until complete? As usual, the answer is that the components of the system evolved earlier, to perform a different function. The enzymes used in recombination are the same enzymes that repair damaged DNA, a process that also requires the precise cutting and splicing of DNA molecules. Meiosis is peculiar to eukaryotes, but most of the enzymes involved were present in prokaryotes, performing a different function.

We think that the theory summarized in Fig. 7.2 is plausible. There are unanswered questions, but the answers may come from a study of existing organisms. In particular, we need to know more about the ecology of organisms with alternating haploid and diploid phases, and about the genetics of the Archaezoa.

Mating types and sexual differentiation

Almost all sexual protists have two 'mating types', + and −. Gametes of the + type will fuse only with −, and vice versa. This has the effect that a gamete will not fuse with another too similar to itself. This is easy to understand. If, as we suggested, the function of gamete fusion is to confer hybrid vigour, fusion of identical gametes, and in particular of gametes that have just arisen by the meiotic division of a single cell, should be avoided. But it is at first sight puzzling that there are just two mating types. Why not many types, any one of which could fuse with any other? This would increase the proportion of gametes with which any particular gamete could fuse. To digress for a moment: something like this happens in flowering plants. Most are hermaphrodite, producing both seeds and pollen, but are self-incompatible: that is, they cannot pollinate themselves. Usually there are many self-incompatible types, and pollen from any one of them can pollinate any of the others. This is just what one would expect. So why only two mating types?

A possible answer is that it is easier to devise a system with two mating types than with many, but evolution is not usually so uninventive. Whatever may have been the case in the first sexual organisms—and two mating types may well be primitive—today the explanation seems to be that mitochondria, and chloroplasts if they are present, are inherited from one parent only. For example, in the well-studied green alga, *Chlamydomonas*, mitochondria are inherited only from the − mating type and chloroplasts only from the + type. This rule of uniparental inheritance of intracellular organelles is almost universal. In animals, mitochondria are maternally inherited. In plants, chloroplasts are usually transmitted in the ovule, not the pollen, but there are exceptions: in conifers, chloroplasts are transmitted in the pollen. It is therefore not surprising that, unlike flowering plants but like mammals, conifers are never parthenogenetic —another example of the difficulty of reverting from sex to parthenogenesis.

Why should uniparental inheritance of organelles be so nearly universal? The likely answer is that, if mitochondria, say, were inherited from both parents, the stage would be set for the evolution of 'selfish' organelles. At cell division, mitochondria are randomly passed to the two daughter cells: there is no device ensuring that one copy of each mitochondrion passes to each daughter cell, as is the case for chromosomes. Therefore a mitochondrion that multiplied more rapidly within the cell, but which was less effective in making ATP, would spread through the population. Such mitochondrial mutants do occur, but because of uniparental inheritance they cannot spread, although they may kill the cell they are in.

Given the need for uniparental inheritance of organelles, it is hard to evolve more than two mating types, one that provides the organelle and the other that does not. In 1992, Laurence Hurst and William Hamilton published a remarkable set of observations showing that this is indeed the right explanation. In ciliated protozoa (for example, the slipper animalcule, *Paramecium*), there is usually no gamete fusion. Instead, two cells lie side by side, or 'conjugate', and each passes a haploid nucleus to the other, without any cytoplasmic mixing. The conjugating cells then separate. Each is a diploid, with a chromosome set from both parents, but each with its own mitochondria. In these ciliates, there are multiple incompatibility types, as expected because there is no risk of the spread of selfish organelles. The proof goes further. In one group of ciliates, the hypotrichs, there are two alternative sexual processes, conjugation and gamete fusion. There are multiple mating types for conjugation, but only two determining gamete fusion. It is pleasing when peculiar and otherwise baffling facts such as these make sense in terms of a theory that was developed in ignorance of them.

In animals and plants, there is a division of labour between a motile gamete and a gamete carrying the food stores needed for the development of a large adult. There are mathematical models showing the circumstances in which such a division of labour would pay: they confirm that adult size is a crucial variable favouring anisogamy. There is again some support for the theory by comparing related isogamous and anisogamous species. *Volvox* is a genus of green algae, related to *Chlamydomonas*, but forming multicellular colonies—hollow spheres of green ciliated cells. In species with small colonies, the gametes are all motile and all the same size. In colonies of intermediate size, the gametes are again motile but of varying size. In species with the largest colonies, there are large non-motile gametes and small motile ones. In *Volvox*, we can see on a small scale the evolutionary path once travelled by the ancestors of the animals and plants. It is interesting that the Bible gets it right. Males were the first sex: females were secondary.

Given anisogamy, secondary sexual characters may evolve. The central logic is that eggs are costly to produce and sperm cheap. Males, therefore, may have more to gain than females by mating repeatedly, This asymmetry often leads to differences in size, weapons, and ornaments between the sexes. This is a grossly oversimplified summary of a complex subject. For example, there are ways of investing in offspring other than putting nutrients into the gametes. If, as in sea horses, males incubate the eggs, females may compete for males rather than the other way round.

To summarize, we can detect three stages in the history of sexual differentiation. First, the evolution of two mating types, + and −, driven by the need for uni-parental inheritance of intracellular organelles. Second, the evolution of males and females, producing motile and food-storing gametes, respectively. Third, the evolution in some lineages of secondary differences between the sexes, driven in part by competition for mates, and in part by a division of labour in raising the young.

CHAPTER 8

GENETIC CONFLICT

The kinds of problem we have pointed to when discussing the major transitions can be observed in present-day organisms. In this chapter and the next, we pause in our chronicle of these transitions to look at examples of conflict and co-operation that can be studied today. We hope in this way to add some realism to the suggestions we have made. In particular, we want to show how difficult it may be to eliminate conflict between replicators within an organism, and yet that remarkable examples of co-operation do in fact exist. In this chapter we describe the conflicts of interest that can arise between genes, or groups of genes, present in the same organism, or in closely related organisms. In the next chapter we discuss examples of co-operation between organisms with different evolutionary histories.

Intragenomic conflict

When we look at an organism, we are struck by the way its parts co-operate to ensure the survival of the whole. This implies that the many thousands of genes in its genome have been programmed by natural selection to co-operate: in current jargon, they are 'coadapted'. Until relatively recently, this result was taken for granted, and cases where the genes in a single organism do not co-operate were seen as rare exceptions. During the past 10 years, attitudes have changed dramatically, as more and more examples of non-co-operation have been discovered. In some cases these discoveries have been more or less accidental. In others—for example, the conflict between mother and fetus described below—they have arisen from a deliberate search inspired by a gene-centred approach to evolution.

Consider first a population of cells in which there is no sexual fusion, and in which a gene present in a cell today can transmit copies of itself only to direct descendants of that cell. The interests of such a gene are identical to the interests of any other gene in the same cell. A gene (or rather, copies of a gene) will survive into the future to exactly the same extent that any other gene in the same cell will survive. If a gene were to undergo a mutation enabling it to multiply more rapidly within the cell than other genes, this would do nothing for its survival:

indeed, it would almost certainly reduce the fitness of the cell, and so reduce the gene's chances of survival. Such a mutation, if it occurred, would be eliminated by selection in just the same way as any other mutation that reduced cell fitness. In such a population, co-operation between genes would indeed be the rule.

But organisms are not asexual. Most eukaryotes have sexual fusion and meiosis. In prokaryotes, although there is no cell fusion, there are many ways in which genes can be transmitted horizontally from one cell to another. Given sex, or the various parasexual processes of bacteria, a gene can increase in frequency, not only by increasing the fitness of the cell in which it finds itself, but by increasing its chances of being transmitted to other cells, even if this reduces the fitness of its current 'host' cell. Co-operation is no longer the inevitable rule. Some examples will make this conclusion clearer.

Some examples of intragenomic conflict

Meiotic drive

Typically, if a diploid individual has two different alleles at a locus, say *A* and *a*, then when that individual produces gametes, half will carry *A* and half *a*. This is what is meant by saying that meiosis is 'fair'. But occasionally meiosis is unfair. For example, in the fruit fly *Drosophila melanogaster* there is a gene known as 'segregation distorter', or *SD* for short. A fly inheriting *SD* from one parent and the normal allele, +, from the other, will transmit *SD* to 95 per cent or more of its offspring. (In fact, *SD* is not a single gene, but two closely linked genes, one producing a toxin that kills gametes not carrying *SD*, and the other protecting gametes with *SD* against the toxin.) Several examples of such meiotic-drive genes are now known. Genes elsewhere in the genome, however, have an interest in suppressing meiotic drive, because drive usually reduces fertility. Thus there is a conflict between meiotic-drive genes and genes elsewhere in the genome. It seems that the rest of the genome usually wins, a fact that led Egbert Leigh to coin the phrase 'parliament of genes'.

Male sterility

We have in fact already described an example of intragenomic conflict (p. 24), in which genes in the mitochondria of hermaphrodite plants cause male sterility, and chromosomal genes suppress their effects and restore male fertility. The important point to notice here is that the conflict arises because the two kinds of gene have different patterns of inheritance, and therefore favour different resource allocation to male and female functions.

Transposable elements

In both prokaryotes and eukaryotes, there are genetic elements in the chromosomes that replicate out of phase with the chromosome as a whole. The molecular mechanisms of this replication are complex and various, and need not bother us here. The result of replication is that an additional copy, or copies, of an element appear elsewhere in the genome, perhaps on a different chromosome. In a strictly asexual organism, such transposition would not increase long-term survival, for reasons already explained. But if there is sex, things are different, because an element inherited by an individual from one parent, if it transposes to new sites in the genome, will be transmitted to more than half the offspring of that individual. In consequence, transposable elements, or 'transposons', can spread rapidly through a population. We know, for example, that a transposon some 3000 bases long, known as the P element, has spread through the species *D. melanogaster* during the past 50 years. It is worth noting that we would not know about this if *D. melanogaster* had not been the subject of extensive genetic study during that period: the spread of new transposable elements may be a commoner event than we know.

Transposable elements of various kinds have been found in all organisms that have been extensively studied, and may constitute more than half the total chromosomal DNA. For example, in humans there is an element known as *Alu,* 282 bases long, present in 300 000–500 000 copies distributed throughout the genome, and accounting for some 5 per cent of the genomic DNA. Related elements are found in other monkeys and apes. So far as is known, these *Alu* elements do nothing useful for the organism. They are only one of many kinds of repeated elements in the human genome. It may seem ironic that, in many organisms, most of the DNA in the chromosomes is there not because it is needed but because a chromosome is a good place for a selfish-DNA element to live and get itself copied; from a gene-centred point of view, however, it is just what we should expect.

Parent–offspring conflict

Meiotic drive, male sterility, and transposable elements are examples of intragenomic conflict between genetic elements within the same individual. In 1974, Robert Trivers explored the conflicts that can arise between the genes in related individuals. For example, genes in siblings, and hence siblings, may compete for resources from their parents. Trivers was particularly interested in conflict between parents and their offspring. At first sight, it might seem that there should be no such conflict. A gene in the mother has a 50 per cent chance of being present in the child, and will therefore be selected to ensure the survival of the child, and so, of course, will genes in the child. So where is the conflict? The answer lies in the '50 per cent chance'. Both genes have an interest in the survival of the child, but they may differ in the degree of sacrifice or resource allocation that should be made to ensure it. In humans, for example, when should a child be weaned, bearing in mind that the mother will probably not become pregnant again until after weaning? A gene in a child might maximize its chances if weaning occurred at, say, 2 years of age, whereas a gene in the mother, which has only a 50 per cent chance of being present in the child, and so has a greater interest in future children, might prefer weaning at 18 months. This could lead to a weaning conflict between mother and child.

When Trivers first put forward these ideas, they seemed to many to be speculative and hard to test. In the last few years, however, the work of David Haig and others has made it very likely that some otherwise unexplained features of pregnancy may best be explained as arising from maternal-fetal conflict. We will mention only one example. A serious risk to the mother during pregnancy is pre-eclampsia, a rise in blood pressure that may damage the kidneys. In typical pregnancies, the fetus is more likely to survive if there is a moderate rise in blood pressure. Fetal cells (or, more precisely, cells in the placenta that are of fetal origin) ensure an increased blood supply to the placenta in two ways: by destroying nerves and muscles in the walls of blood vessels supplying the placenta, and by causing a restriction of maternal arteries supplying other organs. The latter change, in particular, can lead to a rise in maternal blood pressure. In normal pregnancies, this benefits the fetus without damaging the mother, but in exceptional cases it can threaten the mother's life.

Why are we here at all?

The recent flood of evidence demonstrating genetic conflict at various levels, and showing how widespread are selfish genetic elements in the genome, can leave one wondering how complex organisms have managed to survive these many threats to their continued existence. One reason has been discussed already: the fact that most genes in the genome have a common interest in suppressing the ill effects of selfish elements. Two other selective pressures may have helped to prevent selfish elements from taking over.

Lesser fleas

There is some truth in the rhyme:

> Great fleas have little fleas
> Upon their backs to bite 'em,
> And little fleas have lesser fleas,
> And so ad infinitum.

Of course, the sequence cannot go on for ever, but it does stretch further than one might think. For example, the P element of *Drosophila*, discussed above, consists of 3000 bases and codes for only two proteins. One might think that this is about as far as a parasite can be reduced, but if so one would be wrong. Today, the genome of *Drosophila* contains not only functional P elements, but deficient elements that do not code for the enzyme, a 'transposase', that brings about transposition. Such deficient elements are a kind of parasite riding on the back of functional P elements. On their own they cannot replicate, but if there are functional elements in the cell, so that transposase is produced, the deficient elements can transpose to new sites.

Many selfish elements, and also many viruses, are associated with such deficient, parasitic elements, whose presence reduces the ability of the selfish elements themselves to multiply: to that extent, these ultra-parasites help the host organism. Perhaps the ultimate selfish replicator is a circle of DNA consisting simply of a series of short sequences, each of which is a signal to the cell's replicating enzymes, saying 'start replicating here'.

Don't kill the goose

A more important mechanism limiting selfish DNA is selection at the level of the organism itself. A selfish element that is too successful at multiplying within a cell will kill its host, and kill itself at the same time: it has killed the goose that lays the golden eggs. This can cause selfish elements to evolve mechanisms that limit their own growth. The process can also be illustrated by the P element. The element codes for two proteins: the transposase and a regulator protein that limits transposition. In the absence of the regulator, transposition is so frequent that it kills or sterilizes the fly, and in so doing destroys the P elements it carries.

Another possibility that has to be considered is that selection at levels higher than that of the individual organism may be relevant. Species that evolve mechanisms (for example, uniparental inheritance of organelles, inactivation of genes in the male gamete) that rule out certain kinds of selfish genetic behaviour are more likely to survive, as species, than those that lack such mechanisms.

This world of selfish genetic elements and intragenomic conflict is one we are only just beginning to explore. It is bringing home to us just how difficult it has been to evolve complex organisms whose genes co-operate rather than compete.

CHAPTER 9
....

LIVING TOGETHER

One of the themes of this book is that complex organisms depend on a division of labour between their parts. But this end result can evolve in two very different ways. Compare, for example, an elephant and a plant cell. An elephant depends on co-operation between different kinds of cells—epithelial cells, muscle cells, neurones, and so on. These cells have essentially the same genes. They are derived during development by the division of a single fertilized egg. In evolutionary time, they are all descended from the same single-celled ancestor. The differences between them arise not from possessing different genes but because influences external to the cells cause different genes to be active in different cells. The division of labour in human society, or between castes in an insect colony, is analogous. Humans, although not genetically identical, are very similar, and have a recent common ancestor. The differences between a carpenter and an electrician are caused not by their genes but by their training. The evolution of such systems—multicellular organisms, and animal and human societies—is not discussed in this chapter: it is the topic of the remainder of the book.

In contrast, a plant cell depends on co-operation between three genetically different entities, the genomes of the nucleus, the mitochondrion, and the chloroplast. As described in Chapter 6, mitochondria and chloroplasts are thought to be descended from once free-living prokaryotes that were engulfed by a primitive eukaryotic cell. This coming together of once-independent replicators also happened, we think, in the origin of the first cells from populations of replicating molecules.

The process whereby once-independent replicators come to live together in an intimate union is called 'symbiosis'. Although often used to refer to cases in which the partners co-operate, the word properly refers both to cases of 'mutualism', in which the partners do indeed co-operate, and of 'parasitism', in which one partner lives at the expense of the other: 'commensalism' refers to cases in which there is neither co-operation nor exploitation. As we shall see, it is not always easy to decide to which category a particular relationship belongs, and one can readily evolve into another. We first describe some examples, and then discuss the selective mechanisms responsible for their evolution.

The natural history of symbiosis

Symbioses between bacteria and eukaryotes

It is a curious fact that one can kill aphids with antibiotics. Although not recommended as a control procedure for greenfly on one's roses, this sensitivity does tell us that aphids are dependent on bacterial symbionts. Aphids live on the fluids circulating in the plants they attack. This fluid lacks certain substances—in effect, insect vitamins—that the aphids need but cannot synthesize for themselves (in particular, many amino acids). The bacteria help the aphids by synthesizing these substances. They are transmitted to the next generation inside the eggs laid by the aphid: they cannot survive on their own. This symbiosis is ancient: a molecular study of different families of aphids, and their symbionts, has shown that the bacteria have been vertically transmitted, from aphid mother to aphid daughter, for over 50 million years.

In this example, the host benefits because the symbiont can carry out a biochemical synthesis impossible to the host. This is the usual basis of long-term symbiosis between bacteria and eukaryotes. Plants cannot 'fix' nitrogen: that is, they need nitrogen in the form of ammonia or other nitrogenous compounds, and cannot use the molecular nitrogen so abundant in the atmosphere. Leguminous plants, however, form a symbiotic union with a bacterium, *Rhizobium*, that can fix atmospheric nitrogen. It is for this reason that we plant clover in our grasslands, and alfalfa on arable land.

The ecosystem of deep-sea vents depends wholly on symbiosis. Most ecosystems depend ultimately on photosynthesis for their energy. Plants trap sunlight and use the energy to synthesize sugars and other organic compounds: all other organisms in the system depend on plants. The same is true of most deep-sea organisms: it is too dark for photosynthesis so they rely instead on the fallout of dead organisms from the surface layers, which do depend on photosynthesis. But in deep-sea vents there is an alternative source of energy—the sulphides emerging from the vents. The large worm *Riftia* that inhabits these vents has no mouth or anus as an adult, and relies on symbiotic bacteria, housed in a special organ, which oxidize the sulphides. Both oxygen and sulphide are transported to these organs by a special haemoglobin. In effect, deep-sea vents are occupied by a unique ecosystem, not dependent on photosynthesis for its energy but depending instead on symbiosis between animals and sulphur-metabolizing bacteria.

Symbioses between animals and single-celled algae

On the beaches of Brittany, there is a unique flatworm, *Convoluta*. As an adult it resembles *Riftia* in lacking both mouth and anus but it contains symbiotic green algae. When the tide is out, it lies on the surface of the sand, and its symbionts photosynthesize. When shaken by the incoming tide, *Convoluta* burrows beneath the surface, and so avoids being swept out to sea. This has a curious effect: at low tide the sand is green, but if one pats it, it turns golden. *Convoluta* is perhaps a curiosity, but symbiosis between aquatic animals and algae is widespread, and can be ecologically important: the animals that build coral reefs can do so only with the aid of symbiotic dinoflagellate algae.

Symbioses involving fungi

Lichens are perhaps the most familiar example of symbiosis. They are important in colonizing bare rock. A lichen consists of a host fungus containing symbiotic 'algae', which may be either symbiotic green algae or prokaryotic cyanobacteria. The fungi are of many kinds, and it is clear that lichen associations have evolved many times. Almost all the algae found in these associations are found free in nature. The fungi are certainly benefiting from the association, but it is less clear that the algae are getting anything out of it.

Many land plant communities are dependent on an association between mycorrhizal fungi and plant roots. These fungi are known from the Devonian period, when the land was first colonized by plants. At that time, soils would have been mineral in composition, lacking organic nutrients. The fungi were probably important in making minerals available to the plants. Today, mycorrhizal fungi are important in mineral soils, particularly in the tropics. There is a net flow of minerals from fungus to plant, and of organic compounds from plant to fungus, so it seems that both partners are benefiting. A more recently evolved group of mycorrhizal fungi are associated with ericaceous plants (heathers, rhododendrons, and so on) growing on acid soils with high organic content (peat). These ericoid fungi not only supply minerals to the plants but also make available organic compounds that the plants could not acquire without them.

As a final example, we cannot resist describing the symbiosis between leaf-cutting ants and fungi. Leaves, and even flower petals, are cut by the worker ants, and carried to their nest across the floor of the forest, like the banners of a miniature political demonstration. There they are digested by special fungi, farmed by the ants as we farm mushrooms. The fungi are provided with food, and in turn provide food to the ants. This is just the most dramatic of the many ways in which animals that cannot themselves digest cellulose use other organisms to do it for them.

The evolution of symbiosis: mutualism or parasitism?

In the previous section we gave examples in which either both partners are benefiting, or at least the host organism is benefiting. But there are plenty of examples of symbionts that damage or kill their hosts. Can we make any predictions about how evolution will proceed?

Until relatively recently, the conventional wisdom was that symbionts associated for a long time with a given host species would become relatively harmless. The view was well expressed by René Dubos in 1965 in his book *Man adapting*: 'Given enough time a state of peaceful coexistence eventually becomes established between any host and parasite'. On this view, serious disease is caused by parasites invading a new host species for the first time. There is little doubt that this is sometimes true. A recent example is the fact that the human immunodeficiency virus (HIV) is almost always fatal to humans, but the simian equivalent (SIV), from which HIV only very recently evolved, is found in a large proportion of African green monkeys, apparently without causing any harm.

But this is not always the case. For example, typhoid is caused by the bacterium *Salmonella typhi*, which is found only in humans. The related bacterium, *S. typhimurium*, kills its normal host, the mouse, but is harmless in humans. Of course, it may be that in time a state of peaceful coexistence will evolve between these bacteria and their hosts, but the time-scale will be in millions rather than thousands of years.

There are reasons why host–parasite systems should often evolve towards commensalism. Essentially, they are the same as the reasons, discussed in the last chapter, why selfish genetic elements have not destroyed all complex organisms. The host organism will be selected to evolve so as to control the parasite, and the parasite will evolve so as not to destroy a host on whose survival its own future may depend.

That hosts will be selected to reduce the damaging effects of parasites is obvious. Studies of host–parasite systems often reveal traces of a past arms race between them. Thus the main weapon of vertebrate hosts against parasitic microbes is their immune system: they learn to make antibodies against their parasites, in particular against their surface proteins. These proteins evolve much more rapidly than others, to escape immune attack. Some parasites have evolved special mechanisms for periodically replacing the proteins exposed on their surfaces. Of course, such arms races need not lead to commensalism: the parasite may keep one step ahead.

More interesting is the possibility that some parasites may evolve towards commensalism because it pays them to do so. Whether or not this will happen depends, among other things, on the methods whereby parasites reach new hosts. If transmission is vertical (Fig. 9.1a), it is in the interest of the symbiont to keep its host alive and fertile. For example, the bacteria symbiotic in aphids are transmitted only in the aphid eggs. Only those mutations in the bacterium that benefit the aphid will be selected. Transmission of symbionts in the host eggs is unusual, but there are other means of vertical transmission. Termites

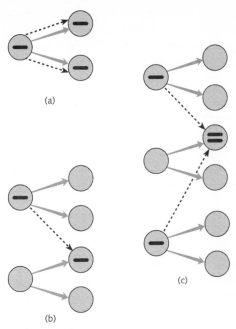

Figure 9.1 Vertical and horizontal transmision of symbionts. (a) Vertical transmission: symbionts are passed directly to all the descendants of the host. (b) Horizontal transmission: a host acquires its symbionts, not from its parent but from unrelated individuals. (c) Horizontal transmission with double infection: a host acquires symbionts from two or more unrelated individuals.

ingest wood, which is digested by a diverse population of symbionts in the termite gut. Larval termites acquire symbionts by licking their mother's anus.

Vertical transmission, however, is not a particularly common feature of mutualistic symbioses. Returning to the examples described earlier, the deep-sea worm *Riftia*, although it lacks a mouth as an adult, has a planktonic larva with a mouth, and acquires sulphur-metabolizing bacteria by swallowing them. The flatworm *Convoluta* also has a mouth when young, and swallows its symbiotic algae. Mycorrhisae in the soil must find plant roots, as must the nitrogen-fixing bacterium *Rhizobium*.

The best-studied example of the evolution of a horizontally transmitted parasite (Fig. 9.1b) concerns the myxoma virus in rabbits. This virus was originally commensal in a South American relative of the rabbit, and was introduced into rabbits as a control measure. Originally, the virus killed rabbits in one or two weeks after infection. Today, the virus only sometimes kills the rabbit, and takes months rather than weeks to do so. It is known that this change has happened because the viruses are less virulent, in addition to the evolution of resistance in the rabbits. The population biologists Robert May and Roy Anderson have analysed this case. They argue that it will pay the parasite not to kill its host, but it will also pay to be highly efficient in transferring to new hosts: it is the product of these two factors, host survival and infectivity, that will be maximized. If the parasite can achieve high infectivity only by damaging the host, we should not expect evolution towards harmless commensalism. Parasites do cause symptoms in their hosts that increase infectivity, as anyone who has suffered from a common cold knows only too well.

There is another reason why parasites may not evolve towards commensalism. Suppose that, typically, a host is infected simultaneously by parasites from two sources (Fig. 9.1c). Then selection on the parasite will favour high infectivity, at the expense of host survival. There is no point in keeping the goose alive if someone else is going to kill it.

Comparing different host–symbiont systems does lend support to the idea that the outcome is affected by the mode of transmission. But the cases of mutualism in which transmission is horizontal are common enough to suggest that the most important factor is the opportunity for mutual benefit.

Co-operation or slavery?

In some cases of symbiosis, it is tempting to ask whether co-operation or slavery is the more appropriate analogy. For example, most of the algae involved in lichen associations are also found free-living. Would it be better to regard them as slaves of the fungus or as willing co-operators? This is not a question that can be answered. The point of analogies of this kind is not that they are true but that they suggest questions to ask, and predictions to test. In this case, we can ask 'are there features of the host organism (the fungus) or of the symbiont (the alga) that have evolved because they help to establish the symbiosis?' In the case of some ancient symbioses—for example, that between eukaryotic cells and their organelles—we cannot say: there is really no way of choosing between the two analogies. But we can do better in some more recent examples. For example, should we regard the fungi farmed by leaf-cutting ants as slaves or partners? The answer seems to be the latter, because the fungi have features that make sense only as attractors for ants: their thread-like hyphae have inflated tips, absent in other fungi, that serve as food for the ants, and seem to serve no other purpose.

Another feature of many mutualistic symbionts is that they have become asexual: few parasites have done this. The explanation, presumably, is that the mutualists do not have to evolve continuously to overcome the defences of their hosts: instead, the host species are selected to make it easy for the symbionts. If this explanation is correct, it has interesting implications. We suggested on pp. 82–3 that there are two rival explanations for the prevalence of sex: first, that it accelerates evolution, and, second, that it reduces the mutational load. The loss of sex by many mutualists suggests that the first of these explanations must be at least part of the truth.

The importance of mutualism

Most of the mutualisms we have described exist because the symbiont can carry out some biochemical process impossible to the host: it can photosynthesize, fix nitrogen, metabolize sulphur, digest cellulose, or synthesize amino acids. Such symbioses have been ecologically important. Today they are the basis of ecosystems in deep-sea vents, coral reefs, tropical forests, and acid moors. Symbiosis between plants and fungi may well have been important in the conquest of dry land.

Symbiosis, then, played a part in three of our major transitions—the origin of the first cells, of chromosomes, and of eukaryotic cells—and in helping host organisms to adapt to difficult environments. It is important, however, not to misunderstand or exaggerate its role. Lynn Margulis, who marshalled the evidence that persuaded biologists that mitochondria and chloroplasts were once symbionts, has sometimes argued that symbiosis is the main source of evolutionary novelty, and that natural selection has been of minor importance. This will not do. Symbiosis is important because both partners contribute something. In nitrogen-fixing symbioses, for example, *Rhizobium* contributes the ability to fix nitrogen, and the plant contributes photosynthesis and the whole anatomy of root and shoot required to succeed on land. These are complex adaptations that could only have evolved by natural selection. The motorbike is a symbiosis between the bicycle and the internal combustion engine. It works fine, if you like that kind of thing, but someone had to invent the bicycle and internal combustion. Symbiosis is not an alternative to natural selection: rather, we require a Darwinian explanation of symbiosis.

The other important point to bear in mind is the one made at the start of this chapter. Not all co-operation between parts arose by symbiosis. In fact, the most complex examples of co-operation between parts specialized for different functions arose by a process of differentiation between genetically identical, or at least similar, entities. We now turn to such processes: the origins of multicellular organisms, and of societies.

CHAPTER 10

THE EVOLUTION OF MANY-CELLED ORGANISMS

An animal's body is composed of many millions of cells, of many different kinds—muscle cells, nerve cells, blood cells of various kinds, and so on. Organisms of this kind have evolved independently on three occasions, giving rise to animals, plants, and fungi. The third of these groups is less complex than the other two, but a mushroom is still a fairly elaborate structure. There are in addition many kinds of simpler multicellular organisms, with only a few types of cells: *Volvox*, a hollow green sphere of ciliated cells, with germinal cells inside, is a charming example.

These many independent origins of a multicellular mode of existence, contrasted with the unique origin, for example, of the genetic code, of eukaryotic cells, and of meiotic sex, suggest that the step may not have been a particularly difficult one. One observation, however, points in the opposite direction. This is the apparently explosive radiation of animals at the beginning of the Cambrian period, some 540 million years ago, suggesting that, once some crucial invention had been made, animals rapidly evolved a range of different body plans, and different ways of moving about, eating, and protecting themselves. This raises two questions. How explosive was the Cambrian explosion? What was the crucial invention, if there was one?

There is no doubt that, in rocks some 540 million years old, there appear for the first time abundant fossils of a great diversity of marine animals. These animals left fossils that we can find because they were large, and had shells or external skeletons that remained when their soft parts rotted away. It is clear that there were few, if any, large animals with shells before the Cambrian, but perhaps there were small soft-bodied animals. We know that such animals can exist without leaving fossils—some existing kinds of animals have left no fossil record. This is true not only of rare and unfamiliar creatures but of some abundant and familiar ones, such as the roundworms (nematodes), which are among the world's most abundant animals.

There is some direct evidence of animals before Cambrian times. Some 20 million years before the Cambrian explosion there are, in rocks from many parts of the world, the remains of a variety of soft-bodied animals, the so-called Ediacaran fauna. Although they had no hard parts to fossilize, they left the impressions of their bodies in the mud when they died. There is debate about how these fossils should be interpreted, but they are thought to include representatives of modern phyla—certainly coelenterates (anemones and jellyfish), and probably annelids (segmented worms), arthropods (animals with jointed limbs), and echinoderms (starfish and sea urchins). But this takes us back only 20 million years before the Cambrian. Attempts to date the origin of the main animal groups by molecular means (using differences between gene sequences to measure time since divergence) suggest that many-celled animals may have originated as much as 1000 million years ago. If so, fossils provide information only about the last half of their history. However that may be, it remains true that something dramatic happened about 540 million years ago: many lineages of animals independently evolved large size, and hard parts.

In the next section, we discuss what biological inventions were needed for multicellular life. We will be left with a puzzle. At least some of the most important mechanisms are found today in single-celled eukaryotes, and even in bacteria. Perhaps no great invention was needed: what was needed was a change in the physical environment making large multicellular animals possible. One suggestion is that, until about 500 million years ago, there was too little oxygen in the atmosphere and dissolved in the ocean to permit such animals to exist. The first animals could not have had a circulatory system, with blood vessels and a pumping heart: such structures would take time to evolve. At first, therefore, oxygen must have reached their tissues by diffusion, a slow process, particularly if the oxygen tension was low. It may be significant that most, if not all, multicellular fossils from the Precambrian era were paper-thin. Such leaf-like animals, and very small animals, could have existed for many millions of years before they gave rise to the larger, hard-shelled animals whose remains mark the beginning of the Cambrian. It is possible that the Cambrian explosion was triggered by a rise in oxygen tension, aided by the emergence of one or more species of predator, whose presence made it necessary for many other animals to evolve the hard shells that constitute the major component of the fossil record.

What had to be invented?

Gene regulation

August Weismann, who first thought clearly about genes—he called them 'ids'—realized that there were two ways in which cell differentiation could come about (Fig. 10.1). One way, which he favoured, was that, when cells divide during development, only some genes pass to the daughter cells: thus only genes needed in the brain pass to future brain cells, only genes needed in the liver pass

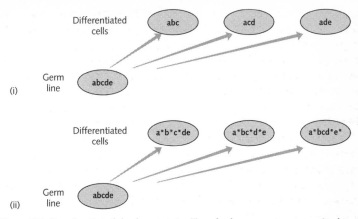

Figure 10.1 Two theories of development. In (i), only those genes are transmitted to a particular tissue that are needed in that tissue: some genes (e.g. *a*) may be needed in all tissues. In (ii), all genes are transmitted to all cells, but different genes are active (indicated by *) in different tissues. August Weismann recognized both possibilities, but thought method (i) more likely: in fact, method (ii) is typical.

to future liver cells, and so on. The germ-line cells, destined to give rise to gametes, are the only cells that retain a complete complement of genes, as they must if they are to be the starting-point of the next generation. But Weismann saw that there was an alternative: all genes pass to all cells, but different genes are active in different cells. To bring this about would require that influences from outside the cell, perhaps from neighbouring cells, should activate the appropriate genes. Weismann rejected this alternative, on the reasonable grounds that he could not see how a sufficiently varied set of outside influences could arise. It is, however, the alternative we now believe to be correct, even though we have not fully solved Weismann's problem of identifying the influences that activate genes.

It is perhaps worth explaining why it is that we now think Weismann's second, less-favoured, alternative is correct. It is not just that it looks as if the process of mitosis is designed to transmit complete sets of genes to each daughter cell. We also know that cells destined to form one structure, and which should therefore, on Weismann's first hypothesis, contain only the genes needed to make that structure, can, if circumstances change, produce a different structure. Every gardener knows that, if one cuts off the tip of a plant shoot and sticks its base into sand, it will often grow roots. This only makes sense if the genes needed to make roots are present in shoots. Weismann was aware of such facts, and of the difficulties they raised for his theory: it was a difficulty he was never able to solve satisfactorily.

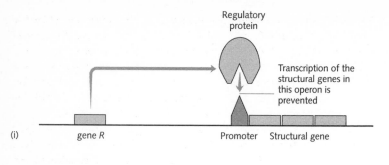

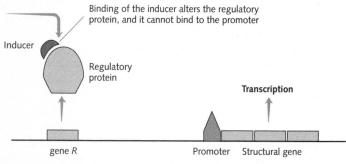

Figure 10.2 The mechanism of gene regulation discovered by François Jacob and Jacques Monod. The gene to be regulated is referred to as a 'structural' gene, to distinguish it from a regulatory gene. (i) A regulatory gene (R) produces a regulatory protein, which binds to a specific promoter sequence at the start of the structural gene, and prevents it being transcribed. In (ii), an inducer binds to the regulatory protein, and alters its shape so that it cannot bind to the protein, thus permitting transcription of the structural gene. By this mechanism, any 'inducer' molecule can switch on any structural gene.

Cell differentiation, then, depends on different genes being active in different cells. An understanding of how this can happen originated in the 1950s in a study by the French biologists François Jacob and Jacques Monod of how the bacterium *Escherichia coli* can acquire the ability to use the sugar lactose. The mechanism is shown in Fig. 10.2. The essential point is that one gene produces a protein, which recognizes and binds to a specific DNA sequence at the start of a second gene (or sometimes of several linked genes), and so is able to regulate the activity of the second gene. In the particular case of the *lac* operon studied by Jacob and Monod, regulation is negative: the regulatory protein switches off the second gene, unless it is rendered ineffective by binding to the 'inducer', lactose. In other cases, however, regulation is positive: the regulated gene is inactive unless it is switched on by the regulator gene.

It turns out that such regulation is a universal feature of living cells. One feature is worth emphasizing. The 'inducer', lactose—which was referred to earlier as an 'outside influence'—binds to the regulator protein at a different site from that at which the regulator binds to the gene it is regulating. A consequence of this, emphasized by Monod in his book, *Chance and necessity*, is that, in principle, any chemical substance can switch on any gene. That is, the 'meaning' of an inducing signal is arbitrary, as the meanings of words are arbitrary. All complex communication depends on such arbitrary signals.

The reader will have noticed that this example of gene regulation comes from a bacterium. In the cells of multicellular animals and plants, genes tend to have many different regulatory sequences, and are affected by many regulatory genes. Hence the activity of a particular gene, in a particular cell, can be under both positive and negative control from different sources, and can depend on the stage of development and of the cell cycle, on the cell's tissue type, on its immediate neighbours, and so on. Gene regulation is complex and hierarchical. Yet the basic mechanism already exists in prokaryotes.

Cell heredity

If you take a few epithelial cells from an animal, and grow them in tissue culture, they will multiply, but they will remain epithelial cells. In the same way, fibroblasts in tissue culture remain fibroblasts, and so do other kinds of differentiated cells. In other words, there is 'cell heredity': like begets like. But the difference between epithelial cells and fibroblasts is not caused by differences between the DNA sequence of their genes: it is caused by a difference in gene activity. There is a dual inheritance system. The familiar system, responsible for hereditary transmission between generations, depends on the copying of DNA sequences during replication. The less-familiar system, responsible for cell heredity, requires the copying, during cell division, of states of gene activation.

The way the second hereditary system works is shown in Fig. 10.3. The activity of a gene is determined by a 'label' attached to the gene: the best-understood labelling system is methylation, but there are others. The crucial point is that, when the cell divides and the DNA is replicated, the pattern of methylation is also copied. When gametes are produced, the labelling pattern must be restored to the initial state: in the language of computer science, a RESET button must be pressed, to restore a default state.

It may come as a surprise that gene activation, determined by specific methylation patterns, is also found in bacteria. Again, we have not discovered an invention that triggered the Cambrian explosion.

Germ line and soma

Early in the development of all but the simplest animals, there is a division into two cell lineages: germ-line cells, which give rise to the gametes—eggs and

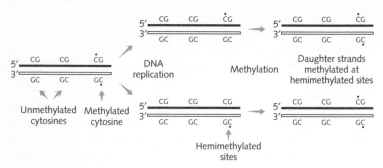

Figure 10.3 Cell heredity in multicellular organisms. The figure shows methylation, the best-understood mechanism. Cytosine (C) is methylated at some G–C doublets: the particular cytosines that are methylated determine the state of gene activity. Cell heredity requires that the pattern is copied when the cell divides. Immediately after DNA replication, the old strand is methylated but the new one is not. An enzyme recognizes these 'hemimethylated' sites, and adds a methyl group to the unmethylated C. There are other mechanisms of cell heredity, depending on protein–DNA binding, that are less well understood.

sperm—and hence to the next generation, and somatic cells that form the rest of the body. For Weismann, this early establishment of a germ line was necessary, because he thought that different somatic cells contained only the genes (ids) needed for that particular tissue: only the germ-line cells contained all the genes, as was necessary if they were to form the next generation. Now that we know that somatic cells carry a complete complement of genes, it is less obvious why such a segregation of the germ line from the soma should occur. It cannot be a necessary feature of the development of complex many-celled organisms, because there is no segregation of a germ line in plants: such segregation is ruled out because there is no way in which germ cells could travel from a central gonad to the flowers, where fertilization takes place. So what advantage do animals gain from the presence of a germ line?

The most likely explanation lies in the nature of cell differentiation and cell heredity. We suggested in the last section that the formation of germ cells requires a process analogous to the pressing of the RESET button on a computer. All the cells must be reset to what has been called a 'totipotent' state: their descendants must be able to differentiate into any of the many kinds of specialized cells of the body. If a gamete were to arise from an epithelial cell, for example, all the gene labelling characteristic of an epithelial cell would have to be undone. As plants demonstrate, there is no reason in principle why this should not happen. But if gametes are formed from undifferentiated germ-line cells, there is less need to change the labels on genes, and hence less opportunity for errors to occur.

Ways of making spatial patterns

At this point, the reader could well object that all this talk of gene regulation misses the real difficulty. How does it come about that the right genes are active in the right places? How does three-dimensional form arise during development?

Before we try to answer this question for animals and plants, it will help to describe three different ways in which non-biological forms can be made. The first mechanism is 'template reproduction'. An example is the production of a pattern by pressing a stamp on a piece of paper. The crucial feature is that a pre-existing form generates a copy of itself by surface-to-surface contact. A second example is the casting of a statue by pouring metal into a mould. No new form is made: a copy is made of a form that already exists. No one thinks that this is the way animal or plant forms are made. Animals are not stamped out, or cast using a mould. The process is crucial in heredity, however: the replication of DNA happens by template reproduction.

A second way of making a pattern is illustrated by such structures as vortices, snowflakes, or the crown of droplets formed when a spherical object is dropped on to water. In these cases, a complex and regular structure is formed by the operation of physical laws. This kind of form is often called 'self-organized'. It arises naturally from the properties of water—incompressibility, viscosity, surface tension, and so on. One might object that the form exists only for an instant, and so cannot be an adequate model for biological form, but this is not a serious objection. There are self-organized patterns that persist: a snowflake is an example. A more serious objection is that, unlike an organism, the water splash does not have parts, or organs, that serve to ensure its survival or reproduction. We argued in the first chapter that adaptedness is a fundamental characteristic of life, so the objection is indeed serious.

A second, related, objection is that the form is not influenced by any informational input: indeed, this is what is implied by the term self-organized. It follows that forms that are strictly self-organized cannot evolve by natural selection, which works by altering the informational input. This is why the water splash does not have organs ensuring its survival. There is a partial way out of this difficulty. The exact shape of the water splash could be altered by changing such things as the density or viscosity of the liquid: that is, by altering what are called the parameters of the system. Some biologists would argue that the best way to picture animal development is as a series of partly self-organized dynamic processes like the water splash, whose parameters are controlled by genes. We think this idea contains an element of truth. For example, it is hard to think that the stripes of a zebra arise in any other way. But we also think that it leaves out a crucial aspect of development. How is it that different genes are active in different cells, at different times, and at different places? We will return to this question. But first we describe a third way of making a non-biological pattern, in which the role of information is more explicit.

A third way of generating a pattern is illustrated by computer graphics. The picture is made by an ink-jet printer attached to a computer. A stream of electric impulses passes from the computer to the printer, and each black spot on the paper is formed in response to one of those impulses. Thus the pattern on the paper is generated by information that was first programmed into the computer, and then transmitted as a stream of electric impulses. There is a one-to-one correspondence between points on the picture and impulses in the wire. Just as no one thinks that biological form is stamped out, so no one thinks that biological form is made as a computer image is made. What is true, however, is that a protein molecule is made in a way analogous to such an image. Thus there is a one-to-one correspondence between amino acids in the protein and base triplets in the gene that coded for it. Change one base and you will change one amino acid. This is not the whole story: the gene specifies the sequence of bases in the protein, but the string must then fold up to produce the three-dimensional form. In most cases, the string will fold up on its own: folding is a self-organized dynamic process, depending on the laws of physics, which do not need to be programmed. (There is a rough analogy between the role of the laws of physics in converting a linear sequence into a three-dimensional form, and the role of the printer in converting a linear stream of impulses into a picture.)

The real objection to a computer image as a model of development, however, is different. Although there is a one-to-one correspondence between triplets of bases in the gene and amino acids, there is no such correspondence between genes and parts of the body. There is not a gene responsible for the nail on your left little finger, and another for the fifteenth eyelash of your right eye. Instead, most structures are influenced by many genes, and most genes influence several structures.

This discussion of three examples of non-biological form may seem rather discouraging. None of them, it seems, is a satisfactory model of development. Yet, when thinking about real development, we find it helpful to have such simple models in mind.

The development of organic form

It is convenient to start with a simple example. In flowering plants, the flower develops from a disc of cells (Fig. 10.4), within which four concentric rings of cells differentiate. The outermost ring gives rise to the sepals, the next to petals, the next to stamens, and the central cells of the disc to carpels. Mutants are known that alter this simple pattern—for example, by converting sepals into

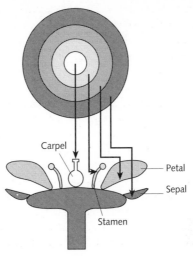

Carpel

Petal

Sepal

Stamen

Figure 10.4 The development of a flower. The flower develops from a disc, formed of four rings of cells, each destined to give rise to different structures.

petals. Some mutants of this kind will be familiar to gardeners: for example, in double paeonies the stamens have been converted to petals. Such mutants have been analysed in the simple crucifer, *Arabidopsis*, with the results shown in Fig. 10.5.

What does this tell us or, more importantly, what does it not tell us? It shows, as we might have guessed, that the genes needed for the development of a particular structure—a petal, say—are activated by particular control genes—genes *a* and *b* in the case of a petal. The genes *a*, *b*, and *c* are at the head of a hierarchy of regulatory genes. But what ensures that these genes are activated in the correct regions of the floral disc, at the right time? In the case of floral development, we do not know. To answer this kind of question, we must turn to *Drosophila* and the mouse, in which studies of developmental genetics have been pursued for longer. The picture becomes horrendously complicated, but a few general principles do seem to be emerging.

As August Weismann saw a century ago, we need a process whereby a group of cells, or a single cell, or ultimately a single gene, is affected by some specific outside influence. One way in which this can happen , called embryonic induction, has been known for a long time. For example, the lens of the vertebrate eye is formed by the differentiation of typical epithelial cells. What makes these cells different from other epithelial cells is that they come into contact with the eye cup, an outgrowth of the developing brain that will become the retina and the optic nerve. Thus a group of cells that would otherwise have become a normal

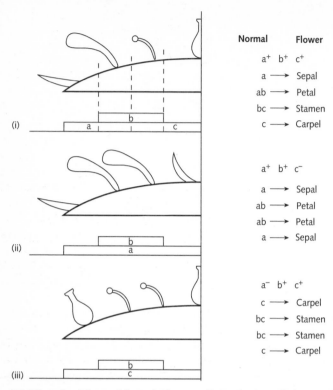

Normal	Flower
a^+ b^+ c^+	
a ⟶	Sepal
ab ⟶	Petal
bc ⟶	Stamen
c ⟶	Carpel
a^+ b^+ c^-	
a ⟶	Sepal
ab ⟶	Petal
ab ⟶	Petal
a ⟶	Sepal
a^- b^+ c^+	
c ⟶	Carpel
bc ⟶	Stamen
bc ⟶	Stamen
c ⟶	Carpel

Figure 10.5 Normal and abnormal flower development. (i), Development of a flower with the normal genotype, $a^+ b^+ c^+$. Below is shown the distribution of gene activity in the flower disc, and above are the structures that develop in response to these activities, following the rules shown on the right. (ii) Development in a flower in which the c gene is mutated and inactive: in the absence of gene c, the a gene is active over the whole disc. (iii) Development in a flower in which the a gene is inactive, and the c gene is active over the whole disc. Other mutants, including double mutants, follow the same rules.

component of the skin are induced to form lens by contact with the eye cup. This has the desirable consequence that the lens forms exactly in front of the retina.

A second mechanism whereby spatial pattern can arise was first suggested 30 years ago by Lewis Wolpert, although its importance has only recently been demonstrated experimentally. First, the theory (Fig. 10.6). Suppose that some chemical substance is produced at a particular point in an embryo. The substance will diffuse outwards, setting up a concentration gradient. Cells locally can respond to the concentration, and different genes can be switched on or off,

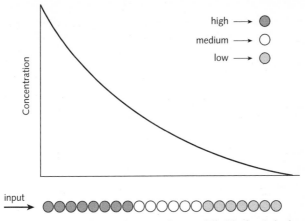

Figure 10.6 The 'French Flag' model of Lewis Wolpert. A diffusible chemical substance, or 'morphogen', is produced at one edge of a sheet of cells, and diffusion sets up a gradient of concentration. Cells respond as shown to the local concentration, forming a pattern with three different regions.

as genes are switched on in *Escherichia coli* by the concentration of lactose. For example, in the egg of *Drosophila*, a concentration gradient is set up of a substance manufactured by cells in the ovary of the female, lying up against the egg at one pole, before it is laid. A high concentration of this substance switches on particular genes in the nuclei of the embryo, and by so doing causes that region of the embryo ultimately to become the head end.

This example is peculiar in that the source of the gradient lies outside the embryo, in the maternal ovary. Typically, gradients arise from within the embryo itself. In principle, a single gradient could specify many embryonic regions, with different concentrations switching on different genes. In practice, however, it seems that not more than two, or at most three, regions are specified by a single gradient: it is interesting that Wolpert used the analogy of the French flag, which has only three regions. It seems that it is easier, or more reliable, for cells to respond to the presence or absence of a substance than to many concentration levels. Thus only a small amount of spatial complexity is generated in a single step. Embryonic development depends on a series of steps, with genes that are switched on in one step being the source of signals in the next.

The evolution of form

Over the years there has been remarkably little conversation between population biologists, who study inheritance and natural selection in contemporary populations, and palaeontologists, who study the fossil record. The former observe genetic change, the latter changes in form. If we do not know how changes in genes cause changes in form, there is not much for the two groups to say to one another. Recent discoveries in developmental genetics may bring this long separation to an end, although it must be admitted that it has not done so yet.

The revolution now in progress in developmental genetics depends on new techniques in molecular biology. It is now possible, in a variety of animals and plants, to identify genes that play a part in early development, to determine their DNA base sequence and the kind of protein they code for, to find out what goes wrong if they are inactivated, to discover where and when they are first active during development, and sometimes to transfer them into distantly related organisms and observe their effects. It is also possible, by looking at their sequence, to work out the evolutionary relationships between genes in different organisms.

By using these techniques, much fascinating information is being accumulated. It is not always easy to interpret, for reasons that can best be explained by an analogy. Imagine you wanted to find out how a motor car engine worked, but all you were allowed to do was to look at particular parts, and to destroy them and see what happened. Suppose that you removed the leads to the sparking-plugs. The engine wouldn't start. The structure of the leads might suggest to you that their function was to carry an electric signal, but not much more. So what would you have learnt? Discovering how a complicated machine works by removing one part at a time and looking at it is not easy. Despite the difficulties, progress is being made, but it is perhaps too early to say what it all means.

One exciting and completely unexpected finding is illustrated in Fig. 10.7. In *Drosophila*, there is a series of genes, known as Hox genes, characterized by the presence, at the start of each gene, of a 'homeobox' domain, coding for 60 amino acids. These genes are active in different regions of the embryo, from front to back. Each seems to act as a master switch, activating a cascade of other genes that are needed for the development of structures appropriate to that region of the embryo. Mutations in these genes cause the appearance of the 'wrong' structures, or, more precisely, of structures in the wrong places. Such 'homeotic' mutations have been known for 50 years, although the genes responsible have only recently been isolated and sequenced. Classic examples are the mutations *antennapedia*, which causes leg-like structures to appear on the head, in place of the antennae, and *tetraptera*, which causes the tiny club-shaped halteres on the last segment of the thorax to be replaced by a second pair of wings.

The unexpected finding is that a similar series of genes exists in the mouse and in other major groups of animals, including annelids and molluscs. The sequence of the homeobox region can be used to discover evolutionary re-

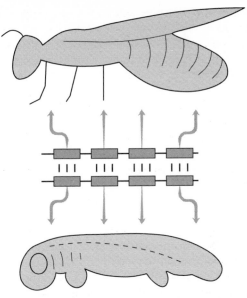

Figure 10.7 The Hox gene family. A series of genes, arranged linearly along the chromosome of the fruit fly *Drosophila*, are active early in development, at different positions along the anteroposterior axis, and induce the development of appropriate structures. A similar series of genes is active in the mouse embryo, and also induce appropriate structures, although the structures that develop in mouse and fly are, of course, quite different. DNA-sequencing studies have shown that the most anteriorly acting gene in the mouse is most similar to the most anteriorly acting gene in the fly, and so on along the series from head to tail.

lationships. It turns out that the most anteriorly acting gene in *Drosophila* is more similar to the anteriorly acting gene in the mouse, and in other animals, than it is to other genes in *Drosophila*, and so on down the sequence from front to back. What this must mean is that the common ancestor of flies, mammals, and segmented worms, and indeed of all bilaterally symmetrical animals, already possessed, some 500 million years ago, a series of Hox genes, acting in different regions of the body from head to tail, and controlling the development of appropriate structures in those regions, and that these genes have been conserved ever since. It has been suggested that it is the possession of these Hox genes that is the defining characteristic of animals, in the same way that the possession of a backbone is the defining characteristic of vertebrates: the characteristic has been called the zootype.

The reason why this discovery is so unexpected is that, although the Hox signalling system has been conserved, there is nothing in common between the structures that it elicits in different groups. For example, there is nothing in mice, or in earthworms, corresponding to the thorax of insects, with its two pairs of wings and six legs. Of course, there is nothing mechanistically puzzling about this: one can use the same kind of switch to turn on a television set or a hair-dryer. What is puzzling is the conservation of the signalling system, despite changes in the structures that it elicits, and so, presumably, in the genes, lower in the hierarchy, that it activates.

This conservation of a signal in evolution is not unique to the Hox genes. An equally surprising example concerns the development of the eye. In the mouse, there is a gene, 'small-eye', which, if it mutates, causes the mouse to be eyeless. Like the Hox genes, it seems to act as a master switch: when it is activated, it switches on a cascade of other genes needed in eye development. A gene with a similar DNA sequence controls eye development in *Drosophila*. The startling finding is that, if the mouse gene is transferred to *Drosophila*, then, wherever it is activated, it causes an eye to appear. The *kind* of eye that develops, of course, is a compound insect eye, not a vertebrate eye. So, when we say that the gene 'controls' eye development, this is rather misleading. What the gene does is to initiate a series of events at a particular site in the embryo, causing an eye to develop at that site: it does not control the kind of eye that develops. In effect, it signals 'make an eye here'. Presumably, the common ancestor of mouse and *Drosophila* had a simple light-sensitive organ, and the position of that organ was specified by activating the gene.

Now that we know about the conservation of signals, it is possible to offer an explanation, although this is being wise after the event. A particular Hox gene in *Drosophila* switches on, not one other gene, but a whole cascade of genes. To alter the signal would alter many features of the resulting structure. Such a drastic change would be most unlikely to improve adaptation: instead, improved adaptation has been achieved by altering, one by one, the genes that respond to an unchanging signal. The conservation of signals follows from the inevitability of gradualism, if adaptive change is to occur.

A second aspect of conservatism in morphological evolution has been recognized for much longer. This is the preservation of a constant body plan, despite changes in way of life. The structural similarity between the human hand, the flipper of a seal, and the wing of a bat, has long been familiar: it is one of the decisive reasons for accepting the theory of evolution. Perhaps more fundamental is the preservation of what has been called the 'phylotype' by members of a given phylum (examples of phyla are the arthropods, chordates, annelids, echinoderms, and molluscs). The point can best be illustrated by the chordates, the phylum to which we belong. During development, all chordates pass through a 'phylotypic stage', the pharyngula, characterized by a notochord (a stiff rod, replaced later in development by the backbone), somites (segmented blocks of tissue destined to give rise to muscles, ribs, etc.), a hollow nerve cord dorsal to the notochord, a pharynx (the anterior part of the alimentary canal, perforated by gill slits, leading from the pharynx to the exterior), and a tail extending behind the anus. At this stage, all chordates are remarkably similar, although later they diverge: adult humans lack notochord, gill slits, and a tail. They may also differ from one another earlier in development: for example, the early embryos of birds, with enormous eggs, and fish such as cod, with small non-yolky eggs, are very different. Development converges on the phylotype, and then diverges.

The first point to be made about the phylotype of chordates is that it echoes the structure and way of life of our earliest ancestors. The notochord, segmented muscles, and post-anal tail were originally adaptations for sinusoidal swimming (as many fish still swim today), and the pharynx was an adaptation for filter-feeding by swallowing sea water and sieving out minute organisms as the water was expelled through the gill slits. It is interesting that some of the basic features of the body plans of other phyla are easily understood as adaptations for particular modes of locomotion: to give a second example, the liquid-filled body cavity and ring-like segments of annelids (for example, earthworms) are adaptations for burrowing.

This may explain why the body plans of different phyla first evolved, but not why they have been preserved. The explanation becomes clearer if we look at development before and after the phylotypic stage. Before, developmental processes are global to the whole embryo and involve extensive cell migrations. In the phylotype, the main body parts are already represented by blocks of undifferentiated cells, arranged relative to one another as they will be arranged in the adult. Subsequent development is more local, with each part developing to some degree independently of the others.

Evolutionary changes in development seem to have occurred in two different ways, for different reasons. Changes early in development, mainly before the phylotypic stage, involve changes in the size and yolkiness of the egg, and the presence or absence of a larval stage. Changes after the phylotypic stage lead to changes in adult structure. It may be that the conservation of the phylotype, as that of the zootype, has been forced on organisms by the need for change to be gradual if it is to be adaptive. It is easier for mutations to cause small morphological changes if the development of different parts is to some degree independent. A genetic programme able to evolve new structures may be one that says, in effect 'set up an unchanging set of parts, with unchanging spatial relationships to one another: then, if you want to change anything, change one part at a time'.

These ideas are vague and speculative. They are in part stimulated by research in a very different area, that of genetic algorithms. Computer scientists are concerned with producing optimal solutions to such problems as the design of robots, railway timetables, and power distribution networks. One way of doing this is by evolution and natural selection. For example, a 'genetic algorithm' that generates a railway timetable is devised, and then allowed to evolve by mutation, recombination, and natural selection. The procedure can work very well, but it depends on devising a way of programming the task such that at least some random changes in the program lead to improvements. This is not easy. As anyone who programs in a familiar language such as BASIC or PASCAL will be aware, random changes lead almost invariably to complete breakdown. Students of genetic algorithms have thought hard about how to make their programs 'evolvable'. The essential feature is that a small change in the program should often result in a small change in its performance: in biological language, a small change in genotype should cause a small change in phenotype (it is intriguing that computer scientists use the terms genotype and phenotype when talking about their programs). One way of achieving this is to make the program 'modular': that is, each part of the program should specify one, and only one, component of the total performance. It is interesting that our bodies are modular—our kidneys, livers, hearts, and legs are separate structures, performing separate functions—and it is beginning to look as if our genetic programme is also to a degree modular. This, at least, is the story that the study of embryos, and the conservation of zootype and phylotype, seems to be telling us.

CHAPTER 11

ANIMAL SOCIETIES

Animal societies with a complex division of labour between the members are of three kinds, and have evolved by three different routes. In this chapter we discuss insect colonies—ants, bees, wasps, and termites—and, more briefly, the colonies formed by some marine invertebrates, which also consist of individuals specialized to perform different roles. In the next chapter we discuss human societies. Apart from the division of labour, and the economic advantages that follow, the three kinds of society have one other thing in common. In all cases, the different kinds of individual are genetically similar: the differences between queen and workers in a bee colony, or between farmer, teacher, and shop-keeper in a human society, arise not from genetics but from the response of similar genotypes to different social circumstances. What a biologist has to explain is how a genotype could have evolved that can respond so differently to different environmental influences.

The existence of non-reproductive castes, the so-called workers, in the social insects, and in some other social animals, poses a formidable problem to the theory of evolution, as Darwin already recognized. Why should worker bees give up reproduction? In what sense does this increase their fitness?

A large part of the answer can be traced back to remarks by Darwin himself and later by J. B. S. Haldane, in 1955: the basic idea was elaborated by William D. Hamilton in the 1960s into a general theory of animal societies. Darwin hinted that, if the family rather than the individual was taken as the target of selection, his theory could be saved. Haldane strengthened this view by saying that he was willing to lay down his life to save two brothers, or 10 cousins. His reason was that these relatives shared, on average, one-half or one-eighth of the genes possessed by him. (Presumably, he said 10 rather than 8 cousins, not because he could not calculate the proportion of genes in common—he was rather good at that kind of calculation—but because he wanted to be ahead on the deal.)

Why should the proportion of shared genes matter? To answer, we have to take a 'gene's-eye view'. A gene that caused Haldane to die, but 10 of his cousins to survive, would cause more genes identical to itself to survive than would a gene that let Haldane live and the cousins die (in fact, 10/8 copies of the gene, on average, would survive, compared with only one that would die). In the same way, genes present in worker bees causing their bearers to give up reproduction for the rearing of their sisters can spread, provided that the advantage of co-operative breeding over individual reproduction is great enough. Thus we are adopting a gene's-eye view, and asking about the conditions under which such 'altruistic' genes can spread. As we shall see, there are genetic systems conducive to the appearance of animal sociality.

The degree of sociality in different species can be placed on a gradient. Most biologists are interested in what is called eusociality—'real sociality'. By definition, eusocial animals must satisfy three criteria:

(1) reproductive division of labour: that is, only some individuals reproduce;

(2) an overlap of generations within the colony; and

(3) co-operative care of the young produced by the breeding individuals.

It is worth noting that, according to this definition, the cells of a multicellular individual are eusocial; we will come back to this issue later. Eusociality is well known in ants, bees, wasps, and termites. It is less well known that a similar degree of eusociality can be observed in naked mole rats, spotted hyenas, African wild dogs, and in some social spiders, notably in the species *Anelosimus eximius*.

Colonies of social animals can, with some justification, be regarded as 'superorganisms', in the sense that they have adaptations (traits increasing fitness) at the colony level. For example, the mound built by termites has a system of air channels that function as an air-conditioning system. On this analogy, the queen and the reproductive males are analogous to the germ line of multicellular organisms, and the non-reproductive individuals would be the soma of the superorganism. Yet one has to be careful with this analogy. There are efficient colonies with several, or even many, queens. In such cases, the relatedness of randomly chosen individuals from a colony is much less than that of randomly chosen somatic cells from an individual. We do not know of any animal species in which different somatic cell lineages, derived from different parents, coexist in a single individual. Such genetic mosaics arise occasionally by chance, and can be produced experimentally, but are never the norm. This observation raises difficulties for the superorganism view.

The eusociality continuum and reproductive skew

When thinking about the evolutionary origin of a phenomenon like eusociality, it is fortunate, as Darwin pointed out, if there exist in the living world examples with a lesser degree of functional specialization, which can give us a hint as to what the intermediate stages may have been. That is why it is much easier to discuss the origin of the eye than of language. Luckily, there does seem to be a 'eusociality continuum', a concept introduced by the behavioural scientist Paul Sherman and his colleagues from Cornell University only a few years ago. Laurent Keller from Lausanne and Nicolas Perrin from Berne have introduced a simple measure, the 'eusociality index', which quantifies the degree of eusociality. It is a measure of the degree to which the direct reproduction of some members of a colony is reduced in favour of altruistic acts directed towards other members of the colony.

When an animal or plant reproduces, it contributes both genes and material to the offspring. This is true for all kinds of reproduction: when a cell divides, it gives not only genes but also cytoplasm and membranes to the offspring cells. In non-eusocial animals, the ratio of transmitted genes and material is the same, barring fluctuations, for all genetically identical individuals, although females typically transmit more material and energy than males. Amongst eusocial animals, some individuals contribute more energy, whereas others contribute more to the gene pool of the next generation, even if they are genetically identical.

In cases of co-operative breeding, we are interested in the way in which 'reproductive skew' varies with the degree of eusociality. If there is skew, some individuals contribute more genes to the offspring than do others. If only some individuals pass on all of the genes, the skew is 1.0; if there is no reproductive division of labour (except between the sexes), the skew is zero. Real animals can be placed rather accurately on this scale. For example, the skew is small in the banded mongoose, in which a group of females raise the young co-operatively; in the smooth-billed ani, a bird in which a group of sisters lay their eggs in the same nest; and in some social spiders. In contrast, many aphid, ant, and termite species have a high skew. The breeders can be regarded as dominant (alpha) and the others as subordinate (beta). In such cases, what makes beta animals accept this situation?

First, it is worth emphasizing that not all is sweetness and light even in the most eusocial of insects. For example, in honey bees some workers have functional ovaries. Since they are not fertilized, the peculiar method of genetic sex determination in bees ensures that they can only produce sons. Yet it turns out that only a small proportion of the drones (reproductive males) in a colony are produced by workers. The reason is that, although workers do lay eggs, these eggs are usually destroyed by other workers. This is exactly what theory would predict: careful calculation shows that a worker has more genes in common with an egg laid by the queen than by another worker (if it would amuse you to check this calculation, you have to know that, in the honey bee, a queen typically mates with many males before founding a colony). This is just one example among many of the conflict that goes on in insect colonies.

Despite such conflict, colonies with reproductive skew do exist. What factors determine the degree of this skew? Behavioural scientists Laurent Keller and Ken Reeve have suggested that the main factors determining the skew are as follows:

1. How successful would a beta animal be if she left the colony and attempted to breed on her own? Clearly, if she could do well on her own, it would pay to leave.

2. How much does co-operation by a beta animal increase the productivity of the colony?

3. What is the structure of genetic relatedness between colony members?

4. What is the chance that a subordinate would win a lethal fight with the dominant without being severely injured?

It is possible to put all these factors into a mathematical model, the details of which need not concern us. By turning the handle of this mathematical device, one can make predictions about the degree of skew to be expected. Fortunately, it is possible to get an intuitive idea of why the predictions come out the way they do. It is also pleasing that the predictions fit with what happens in the world. We will give one example of this fit between theory and observation.

Consider the notion of 'staying incentive'. The dominant female benefits if she can persuade the subordinates to stay and help her. To do this, she may have to permit a certain degree of direct reproduction by the subordinates; if she doesn't, they will leave. But how much reproduction should she allow? This depends on circumstances. If relatedness is high, the staying incentive offered by the dominant can be less than otherwise, because a subordinate (or, more precisely, a gene in a subordinate) has more to gain by staying. If the subordinate is likely to die, or fail to breed, if she leaves, then the incentive for staying can be small. These predictions work out very well. Sometimes, extraordinarily detailed predictions are possible. For example, in dwarf mongooses dominant individuals do not completely suppress reproduction by the subordinates. The chance that a subordinate will be able to breed if she leaves the colony increases with age. The model predicts that older subordinates should be offered a higher staying incentive by dominants than younger ones. This is exactly what happens: older subordinates reproduce more.

The skew model also helps to explain the evolution of morphological castes. In species with a morphological difference between castes (for example, between workers and soldiers, or between workers specialized for different tasks), caste membership determines reproductive contribution. The skew model therefore predicts that morphological castes should be found in species in which ecological constraints are strong (solitary reproduction is very difficult), relatedness is high, and subordinates are unlikely to win fights. Workers are, as expected, small relative to the queen.

Kin selection and genetic relatedness

Genetic relatedness plays a crucial role in the transition from solitary to social breeding. The basic theory for this was put forward by William Hamilton, now in Oxford, more than 30 years ago. In order to account for the spread of altruistic genes in populations, he introduced the novel concept of 'inclusive fitness'. The basic idea is simple: when contemplating the spread of a gene causing its bearer to behave in an altruistic manner, one must consider not only its (adverse) effect on its bearer, but also its (beneficial) effect on the bearer's relatives, multiplied by the respective degree of genetic relatedness. If one gives up direct reproduction for reproduction by one's brothers, for example, one must be able to increase their fitness to such a degree that the expected spread of one's own genes becomes more effective than through the direct route: bearing in mind that brothers share, on average, only half their genes, this implies that, in J. B. S. Haldane's words, one should lay down one's life for (just more than) two brothers. Haldane assumed, however, that the altruistic gene was rare in the population: it is a lot harder to see what will happen if the gene is not rare. What Hamilton did was to show that the same ratio holds true when the altruistic gene is already common in the population, to extend the argument to relatedness in general, and to use the theory as the basis for a general explanation of sociality.

Multiple queens

Having introduced the skew and kin-selection models, we must return to the paradox of multiple queens. Genetic relatedness is usually high in co-operatively breeding birds and mammals. Workers of social insects are closely related to the brood that they look after, provided those offspring, and the workers, are offspring of a single queen. But if there are several queens, Hamilton's framework seems to break down. And there are such colonies: they are called polygynous ('gyn' = female in Greek). The average number of queens per nest can be as high as 100. The situation would not be so hard to explain if the queens were related, but there are cases in which queen–queen relatedness is effectively zero. Given multiple queens, then, one would expect conflict between queens and workers, and between the workers, leading to the collapse of the colony.

The explanation of this apparent paradox may lie in a process that we have already invoked when explaining the maintenance of sexual reproduction. We then talked of 'sexual hang-ups': when a lineage has been reproducing sexually for many millions of years, various secondary adaptations may come to depend on the sexual process, so that it becomes impossible for sex to be abandoned. The same may be true of the social insects. Species with multiple queens have highly complex societies. It is simply not an option for a single female to establish a successful colony on her own: new colonies are established by groups with both queens and workers. But, one might ask, why do not workers lay male eggs? We pointed out above that this is prevented because workers prefer to raise eggs laid by the queen, and so destroy eggs laid by other workers, but this argument breaks down if relatedness is low enough. It turns out that workers in multi-queen colonies are often totally sterile, a rare phenomenon in single-queen species. It looks as if worker sterility has favoured the later evolution of multiple queens.

Even if one explains the maintenance of sociality despite the presence of multiple queens as a 'social hang-up'—that is, that sociality has become so complex that there is no way back—one certainly cannot explain the origin of sociality in this way. No doubt, in the first eusocial colonies there was a single queen, and relatedness was crucial. There also had to be ecological circumstances that made co-operation pay. In the Hymenoptera, it is thought that communal nesting— that is, females breeding independently, but as close neighbours—may have been the first step towards eusociality.

The advantage of such communal breeding arose because females had to forage for food at a distance from the nest. In existing solitary bees and wasps, the major cause of mortality among the juvenile stages is attack by parasitoids— insects that enter the nest and lay their eggs in or on the bodies of the host insect. In a community of breeding females, some could defend the nests while others foraged. This may have been the ecological feature that favoured the evolution of sociality in the Hymenoptera. In the case of termites, the crucial preadaptation may have been their peculiar way of feeding. In order to be able to digest wood, young termites must acquire specialized flagellates, which live in their gut. This they do by feeding at the anus of an adult. This requires close contact between at least two individuals—initially, between mother and offspring. It was maternal care that was the first step towards termite sociality.

The paradox of indiscriminate altruism

If individuals are selected to behave altruistically towards their relatives, one would expect that animals would be able to distinguish relatives from non-relatives. This is not always the case. For example, if baby mice are transferred between broods, the fathers do not discriminate between their own young and others. As Martin Daly and Margo Wilson, two American students of animal behaviour who have increasingly turned their attention to evolutionary explanations of human behaviour, have pointed out, however, this is not really surprising. Animals that nest in a hole or burrow—and mice are such a species —do not need to recognize their own young individually; any young in their own nest is likely to be their own offspring. In contrast, animals whose young are born in the open, and are mobile at or soon after birth, are very good at knowing who is theirs. Yet there are cases in which animals fail to recognize their relatives when it might pay them to do so. For example, when there are several queens in a colony, workers do not always discriminate between their siblings and offspring of other queens. Why should this be so?

Obviously, nepotism depends on the efficiency of recognition: if you cannot recognize your relatives from non-relatives, you cannot act nepotistically. The efficiency of recognition has to be high in the following situations:

- the benefit of nepotism to the recipient is low
- if recognition fails, there is a high cost to relatives that are discriminated against
- the cost of recognition is high, and
- the difference in relatedness of the potential altruists and the individuals it might help is small.

Bearing these points in mind, there are two reasons why, in particular cases, nepotism may be absent. The first is that the advantages of helping close relatives is more than offset by the cost of neglecting less-related individuals. For example, it could be that inspection of relatedness may be so costly in itself (in terms of time, for example) that it does not pay to be nepotistic. This may explain why workers with different fathers and the same multiply-mated queen as mother do not discriminate against one another. Note, however, that this does require some relatedness between the altruist and at least some of the potential recipients.

A second explanation is that kin recognition may be absent because of the cost of recognition errors. Consider, for example, a male bird with a brood, only some of which are in fact his offspring. If, in error, he feeds one of the brood that is not his own, there is a cost, but it is not all that great: if, also in error, he fails to feed one of his own offspring, that offspring would die, and the cost to the father would be high. The ability to recognize kin would not evolve because inefficient recognition would do more harm than good.

What of altruism in human societies? It is obvious that the amount of help we are prepared to give to others often depends on relatedness. The almost universal myth of the wicked stepmother illustrates this fact. There is also clear evidence that neglect and abuse by step-parents are commoner than by biological parents. But there is no genetic inevitability about this. Many step-parents are good parents, and one cannot explain this by saying that they are unaware of their relatedness, or otherwise, to particular children. Humans are often altruistic to others to whom they are unrelated. Extreme examples include voluntary celibacy, and foolhardy bravery. Attempts have been made to explain such behaviour in terms of inclusive fitness: for example, perhaps a celibate monk more than makes up for his own lack of direct reproduction by the help he can give to relatives, and perhaps the risk of death if one is foolishly brave is compensated by sexual success if one is lucky enough to survive.

In general, we do not find such explanations convincing. We do not doubt that inclusive fitness theory will explain a lot about human behaviour, but we are equally convinced that culturally acquired beliefs are important. For biologists, the question that must be answered is why it should be that humans can so readily be influenced by myth and ritual to do things that do not increase their inclusive fitness: this is a problem to which we return in the next chapter.

The division of labour in insect colonies

A reproductive division of labour between fertile queens and sterile workers is a defining feature of eusociality. It has an obvious similarity to the division between germ line and soma in an individual organism. We now turn to a second kind of division of labour that resembles the differentiation between different cell types within the soma: this is the division of labour between workers doing different tasks. Such division of labour can be achieved in two ways. First, unspecialized individuals can be busy at different tasks at the same time: for example, some may be foraging and some may be caring for the brood. The proportion of individuals engaged in different tasks varies according to the colony's needs, and we have to understand how this task allocation comes about. In other cases, there is a morphological difference between workers performing different tasks. The difference may be only a matter of size, with the larger workers defending the nest and the smaller ones foraging. In other cases, the soldier caste may have special weapons. An extreme example of morphological differentiation occurs in the honeypot ants, in a special caste of workers with enormously expanded abdomens functioning as jars in which honey is stored.

How does it come about that an insect colony has workers performing different tasks in the right proportions? It clearly would not do if the colony contained only workers caring for the brood but no foragers, or only soldiers and no workers. First, consider the division of labour in a colony without morphological differentiation. One way in which an appropriate division of labour is achieved is through adults of different ages performing different tasks. For example, young honey bee workers work in the nest, and older bees forage. This mechanism, which has no close analogue in cell differentiation, ensures that all types of tasks are performed, but is less able to ensure that the numbers of workers performing different tasks will change appropriately as the demands on the colony change.

Often, workers of the same age, and the same morphology, can perform two or more different tasks, although of course not at the same time. This allows for flexible adjustment in task allocation in response to changes in conditions. In effect, workers adjust their behaviour to environmental cues. For example, whether or not a honey bee will forage depends on the amount of nectar in the hive. The evidence for such readjusting mechanisms comes from perturbation experiments; for example, extra food can be provided and the response of the colony recorded.

To ensure orderly switching between tasks, environmental cues are not enough: individual workers must communicate with one another. There is no means of global communication within the nest. The image of a queen telling the workers what to do is misleading: bees and ants do not have radio broadcasts to tell them what to do. A better image is the orderly growth of an individual body, brought about by communication between neighbouring cells. Work in the colony is organized by local communication between individuals. The effectiveness of this signalling can be astonishing. One of the most remarkable discoveries of this century was the description by Karl von Frisch of the way in which a honey-bee worker that has discovered a source of food can convey to other bees the nature of the food and, by dancing, the distance and direction of the source.

Properly to understand the dynamics of task allocation and colony welfare, one must build models of the process, as has been done by Deborah Gordon and her colleagues at Stanford University. The models depend entirely on interactions between individuals, and show how global order can result from local rules. There is a real analogy here with the way in which, in development, morphological form appears from local interaction between cells, without the need for any one cell to have an image of the final result.

We turn now to colonies consisting of morphologically differentiated castes. It is helpful to compare this to the differentiation of cell types within the body. All the cells in your body have the same genetic constitution: the differences between them are caused by a secondary (epigenetic) inheritance system. The members of the various castes in an insect colony are also genetically similar, and different because they have been exposed during development to different environments. They are induced by different diets, and by different chemical signals, to develop in different ways. Unlike the cells in an individual body, however, the workers in an insect colony are not genetically identical. This raises the possibility that different genotypes in a colony may be predisposed to develop into morphologically distinct castes. This could at best be a predisposition, not a genetically determined fate: otherwise, colonies might contain only workers, or only soldiers. As yet, the role of genetic predisposition in insect colonies is a matter of speculation rather than solid knowledge.

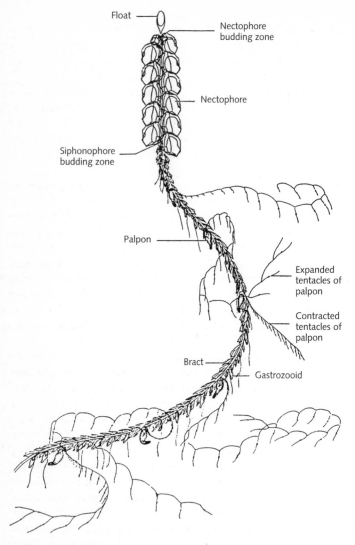

Float

Nectophore
budding zone

Nectophore

Siphonophore
budding zone

Palpon

Expanded
tentacles of
palpon

Contracted
tentacles of
palpon

Bract

Gastrozooid

Figure 11.1 The colonial medusa, *Nanomia cara*. The float provides buoyancy, the nectophores propel the colony, the palpons and gastrozooids capture and digest prey, and the bracts help to protect the colony against physical damage. Sexual zooids produce gametes that unite by conventional fertilization.

Colonial hydroids: organisms evolved from colonies

We turn now to a very different kind of colony. When we look at an ant colony, the individual ants are obvious: with some insight, we can also see that the colony as a whole has some of the characteristics of an organism. But now consider the animal shown in Fig. 11.1. *Nanomia* is a coelenterate, a relative of sea anemones and jellyfish. It looks like a single organism, with a bladder to keep it afloat, pumps to propel it through the water, tentacles for killing prey, digestive organs, and organs for producing gametes. Yet all of these different structures turn out to be modified individuals, or zooids. *Nanomia* is a colony of highly differentiated individuals.

The ancestors of *Nanomia* were individual animals resembling the present-day *Hydra*, a small, cylindrical aquatic animal with a mouth but no anus, and a ring of stinging tentacles surrounding the mouth, feeding on small crustaceans such as *Daphnia*. *Hydra* itself lives in fresh water, but most hydroids are marine. Many of them form simple colonies of similar feeding individuals, connected by a root-like stolon. Such a colony develops from a single fertilized egg. As the stolon extends over the rock to which it is attached, new hydra-like individuals are budded off. *Nanomia*, which is no longer attached to a substrate, also develops from a single fertilized egg, but the individual zooids are differentiated to perform different functions—float, pump, tentacle, stomach, and gamete-producer.

This is a new way of forming a complex body, different from that adopted, for example, by insects and vertebrates. During evolution, and in individual development, insects and vertebrates form a series of initially rather similar segments, which later become differentiated. There is a fundamental difference between this way of forming a complex body, and that adopted by *Nanomia,* and other colonial coelenterates. The segments, and other body organs, of higher animals were never free-living individuals, whereas the body parts of *Nanomia* have evolved from once free-living individuals. *Nanomia* is a colony, but an insect is not.

Although *Nanomia* is a colony, it differs crucially from a colony of ants or termites. *Nanomia* develops from a single fertilized egg. If the parts of its body are more closely integrated than are the individuals in an insect colony, this, as Edward O. Wilson pointed out, is because their relatedness is maximal—that is, unity.

CHAPTER 12

FROM ANIMAL SOCIETIES TO HUMAN SOCIETIES

Despite the obvious similarities between a termite mound and a human city, there are profound differences between the mechanisms that lead to co-operation in the two cases. One important feature of human societies—namely the recognition of individuals—already exists in some social mammals and birds. Although insects may recognize group membership, they do not recognize individuals. In contrast, a monkey recognizes other members of its troop as individuals, and behaves differently towards them. As the phrase 'pecking order' implies, the members of a flock of chickens sort themselves into a linear dominance hierarchy: this probably requires individual recognition. Students of baboons and other monkeys have observed the formation of alliances, in which two or more individuals support one another in conflicts with other members of the group. Such alliances may be based on genetic relatedness, but this is not always so.

The essential points are that, in higher animals, social interactions within a group depend on individual recognition, and that an individual's behaviour towards another depends both on genetic relatedness, and on a memory of previous interactions with that individual.

The characteristics of human societies

It is often said that the defining characteristic of human societies is cultural inheritance; that is, individuals in a society acquire their beliefs and behaviour, their knowledge and skills, by learning from previous generations, and not by genetic inheritance. There is obviously a lot of truth in this idea, particularly when talking of the differences between one individual and another, or between one society and another. At the level of the individual, differences in political opinions are not caused by differences between genes. At the level of societies, the differences between the inhabitants of London today and in the year 1098, or between the inhabitants of London and Beijing today, are caused culturally and not genetically. Having said this, however, there are some reservations that need to be made. First, there is some cultural inheritance in animals, a fact that is important when thinking about the origins of human culture. Second, the ability of humans to learn, and to build societies dependent on cultural transmission, is genetic: human societies differ from chimpanzee societies because humans and chimps differ genetically. Third, humans learn some things more readily than others: the human mind is not a blank slate upon which experience can write what it will.

Young rats can acquire a preference for a new food by smelling it on the coat of other rats. This is a kind of cultural inheritance: two groups of rats feed on different foods, and the difference is transmitted by learning. The mechanism has been called 'local enhancement': the adults create an environment in which it is easier for the young to learn. This contrasts with 'observational learning', in which one animal watches what another is doing, and copies it. For example, in Britain, where milk is sometimes delivered to the door in bottles with shiny metal tops, great tits learn to tear open the tops to get at the cream. The habit is culturally transmitted, as is shown by the fact that in some areas, at some times, tits do not do it. One might guess that the mechanism of transmission is observational learning: young tits see their elders opening bottles and copy them. But this is probably not so. Most animals seem incapable of observational learning. The explanation of bottle-opening seems to be that young birds, in flocks that have the habit, encounter bottles with torn tops, and so learn that there is cream to be found by tearing them: they do not watch other tits opening bottles and copy them.

But it is hard to believe that all culturally inherited traits in animals depend only on local enhancement. For example, in some areas of Greece, golden eagles feed largely on tortoises. They are unable to break open the shell with their beaks, so a bird picks up a tortoise, flies up to a considerable height, and drops it onto the rocks below, thus breaking the shell. It would be absurd to suggest that in Greece, but nowhere else, this behaviour in eagles is genetically programmed. A young bird could learn by local enhancement that tortoise shells contain meat that is good to eat. But how, other than by copying, could they learn to fly up carrying a tortoise, and drop it? We give a second example, from chimpanzees, below.

The distinction between local enhancement and observational learning is important, because only observational learning can lead to cumulative cultural change, which is the characteristic feature of human history. By observational learning, young individuals can learn from adults, but also, if one individual stumbles on an improved way of doing something, that improvement can be copied. The result is that change can be continuing, rather than occasional, and that an individual can learn, by copying, a skill that it could never have learnt on its own.

It is clear that humans depend on observational learning, reinforced by teaching, including verbal instruction. As the example of golden eagles shows, observational learning is not unknown in animals. The Swiss zoologist Christophe Boesch gives the following example in chimpanzees. Some, but not all, populations of chimpanzees dip sticks into the nests of driver ants, and feed on the ants that crawl up the sticks. The chimpanzees in Gombe National Park in Tanzania use a different technique from those in the Taï National Park in Côte d'Ivoire, and catch about four times as many ants per minute. Local enhancement could explain why some populations dip for ants whereas others do not, but cannot explain why the Taï chimpanzees continue to use an inefficient technique, when there is no reason why they should not adopt a more efficient one. But continued use of an inefficient technique is what we expect if youngsters copy their elders.

If higher animals, at least sometimes, are able to copy their elders, why is it that continuous cultural change does not occur among them, as it does in humans? Thus one population of chimpanzees may differ from another for cultural reasons, but a given population is not continuously acquiring new habits. The likely explanation is that, in humans, the main mechanism whereby culture is transmitted is language. The nature and origin of language are discussed in the next chapter. At the risk of repetition, two conclusions of that discussion will be summarized here: they are the close analogy between genetic and linguistic methods of transmitting information, and the implications of linguistics for the modular nature of the human mind.

Both the genetic and linguistic systems are able to transmit an indefinitely large number of messages by the linear sequence of a small number of distinct units. In genetics, the sequence of four bases enable the specification of a large number of proteins, and these, by their interactions, can specify an indefinitely large number of morphologies. In language, the sequence of some 30 or 40 distinct unit sounds, or phonemes, specify many words, and the arrangement of these words in grammatical sentences can convey an indefinitely large number of meanings.

Richard Dawkins has emphasized this analogy by introducing the concept of a 'meme', the unit of cultural inheritance analogous to a gene. A meme, he argues, is a replicator. If we invent and tell you a limerick , you may tell it to your friends, and they to theirs: a single original entity—the representation of the limerick in my brain—has replicated, as a gene might replicate. Clearly, there is room for selection: if we invent a funny limerick, it is more likely to replicate than if we invent a boring one. Of course, whether a meme will replicate, or fail, depends on the nature of the human mind, and on the cultural milieu (that is, on the other memes present in the population). But the same is true of a gene: its increase depends on the environment and on what other genes are present.

There are, of course, differences. Genes are transmitted from parent to offspring: memes can be transmitted horizontally, or even from offspring to parent. But there is a deeper difference between genes and memes. Genes specify structures or behaviours—that is, phenotypes—during development: in inheritance, the phenotype dies and only the genotype is transmitted. The transmission of memes is quite different. A meme is in effect a phenotype: the analogue of the genotype is the neural structure in the brain that specifies that meme. When I tell you a limerick, it is the phenotype that is transmitted: I do not pass you a piece of my brain. It follows that, in the inheritance of memes but not of genes, acquired characters can be inherited. If I tell you a limerick and you think of an improvement, you can incorporate it before you pass it on. In this sense, cultural inheritance is Lamarckian. For these reasons, one cannot readily apply population genetic theory to cultural inheritance. But the analogy between memes and genes can be suggestive in a qualitative if not in a quantitative sense. Further, although for simplicity we have illustrated the idea of a meme by the example of a limerick, it can refer to more important examples, like a belief in the Trinity, or a knowledge of how to manufacture gunpowder.

A second implication of linguistics is the notion of a modular mind. A study of language and its acquisition suggests that the ability to speak is not an aspect of general intelligence, but is a special competence. As Noam Chomsky has argued, we have a special 'language organ'. The evidence for this view is discussed in the next chapter. It has led to the suggestion that the brain may have other domain-specific competences: in current jargon, that the brain may be 'modular'. We return to this idea later in this chapter.

From ape to human

All Old World monkeys and apes live in social groups, with the single exception of the orang-utan. Figure 12.1 shows a reconstructed phylogeny, or ancestral tree, of these animals, and the nature of their social structure. Old World monkey females remain in the social group in which they were born: males leave their natal group before sexual maturity, and must enter another group to breed. They are said to be 'female kin-bonded'. In chimpanzees, the situation is reversed: males remain in their natal groups, and females move. As shown in the figure, other hominoids vary in their social systems, but none is female kin-bonded. Robert Foley, from whom Fig. 12.1 is borrowed, argues that male kin-bonding originated in the common ancestor of humans and chimpanzees. This is the most parsimonious assumption, given the phylogenetic tree shown, with humans and chimpanzees more closely related than either is to gorillas. If so, male kin-bonding is the ancestral condition for members of the human family —hominids. The social systems of modern humans are so varied that it is hard

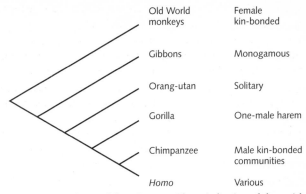

Figure 12.1 An ancestral tree of the primates, with an indication of the social systems associated with different groups, after Robert Foley. The social systems of humans are so varied that it is hard to know what system was characteristic of our hominid ancestors, but Foley suggests that male kin-bonding may have originated in the common ancestor of humans and chimpanzees.

to be sure that this conclusion is correct, but it is the best we can do from the comparative evidence.

The fossil record provides a second source of information about human origins. This is summarized in Fig. 12.2. It is illuminating to compare this record with what is known of human technical achievements, if only for the puzzles that the comparison raises. The australopithecines were bipedal, and lived in wooded grassland. Their relative brain size was only slightly larger than that of apes, and their tool kit was limited and uninventive. In the lineage from *Australopithecus* through *Homo habilis* and *H. erectus* there was a gradual increase in brain size, but rather little technical innovation. The most advanced tool used by *H. erectus* was the handaxe, made from a single block of stone worked on both surfaces, and symmetrical in shape. Such handaxes first appear some 1.4 million years ago, and persist almost unchanged for over a million years: hardly an example of cumulative cultural change.

The most rapid increase in relative brain size has occurred in the past 300 000 years, culminating in the appearance of effectively modern humans some 100 000 years ago. Yet the acceleration in human technical inventiveness, with the appearance of a varied range of tools made of stone, bone, and antler, dates back only 40 000–50 000 years. Burial of the dead, art in the form of cave painting and musical instruments, personal adornment, and trade, originated at much the same time. From about 40 000 years ago, we are faced with evidence of continuing cultural innovation. This raises several problems. Why the delay of 50 000 years between the appearance of the first anatomically modern

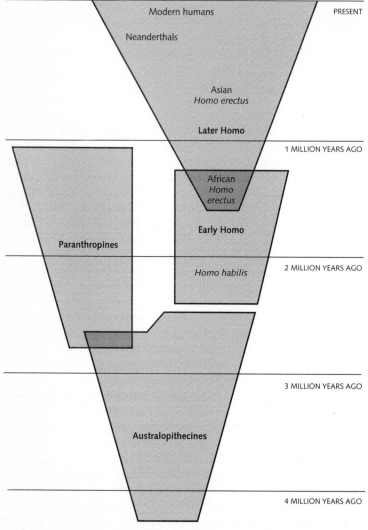

Figure 12.2 The human fossil record, after Chris Stringer and Clive Gamble. Four grades are tentatively recognized. Australopithecines were early, small-brained but bipedal hominids. There was then a division into two main ecological types: heavily built and mainly vegetarian paranthropines, and a more slenderly built lineage, *Homo*, showing an increase in brain size, and leading to Neanderthals and modern humans. The distinction between 'gracile' and 'robust' forms already existed among the later australopithecines. Early *Homo* and *Paranthropus* coexisted for at least 1 million years. There is still debate about the number of coexisting species within each lineage.

humans and the technological revolution? What selective force was responsible for the accelerated increase in brain size 300 000 years ago? When and why did language as we know it originate?

The problems are difficult, because a fossil skull can tell us rather little about the brain that was once inside it, and stone tools little about the society that made them. In *The prehistory of the mind,* published in 1966, British archaeologist Steven Mithen attempted an answer. Although speculative, his book does combine information from palaeontology, archaeology, and psychology to give a convincing answer to these questions. The essence of his argument is as follows. The human mind does indeed contain modules adapted to particular tasks, as suggested by studies of linguistic competence. During much of human evolution, these modules increased in efficiency but remained to a large degree isolated from one another. Language evolved in the first instance to serve social functions, but once grammatical competence had evolved, it provided a means whereby the barriers between modules could be broken down. The burst of creativity during the past 50 000 years resulted from the breaking of these barriers.

Mithen supposes the existence of three mental modules concerned, respectively, with social intelligence, with technical intelligence, and with natural history —that is, with the knowledge of animals and plants needed for efficient foraging. We discuss these in turn.

Social intelligence is a common characteristic of primates. Robin Dunbar, an anthropologist based in Liverpool, has argued that it is the main reason for the increase in brain size in monkeys and apes: his evidence is that there is a striking correlation between brain size in a species, and the size of social groups in that species. A crucial question is how far apes and monkeys have what has been called a 'theory of mind'. To have a theory of mind is to be able to ascribe to others the possession of a mind like one's own, with similar desires and powers of reasoning. There is no convincing evidence that monkeys have such an ability. For example, a vervet monkey gives a different alarm call if it sees an eagle, a snake, or a leopard. But American zoologists Dorothy Cheney and Robert Seyfarth, now at the University of Pennsylvania, who studied these monkeys in the wild, argue that the monkey does not have in mind the knowledge that another monkey may hear its call and respond appropriately; for example, a monkey may continue to call after all others have responded. The behaviour of chimpanzees has convinced most of those who have studied their social behaviour, and skill in manipulation and deceit, that they do indeed have a theory of mind. We can conclude that selection for social intelligence was a major cause of the increase in brain size in monkeys, apes, and humans, and that a theory of mind was already present in the common ancestor of chimpanzees and humans, some 5 million years ago.

Chimpanzees do use tools in the wild; for example, some populations use stones to crack nuts. Even in captivity, however, their ability to make tools is very rudimentary. Australopithecines used tools, but there is no convincing evidence for deliberate toolmaking. The latter skill first appears associated with the remains of *Homo habilis*, but the tools are little more than irregular chipped stones. *Homo erectus* marks a clear advance, with the manufacture of symmetric handaxes, indicating that the toolmaker had an image of the desired result in mind, and the skill to realize it. But, as mentioned earlier, there remains an astonishing conservatism. Thus there is evidence of a limited increase in technical intelligence, combined with a lack of inventiveness.

There is also evidence for a degree of independence of social and technical intelligence even in modern humans. For example, students of autism have suggested that autistic children are deficient in understanding the behaviour of other humans—what has been called 'folk psychology'—but better than average in understanding the behaviour of inanimate objects—'folk physics'.

Finally, there was obviously selection for improved foraging skills, and hence for knowledge of animals and plants, and of their distribution and behaviour. But was this achieved by an increase in general-purpose intelligence, or by the evolution of a specialized natural history module? In favour of the latter, it has been argued that all human societies share certain ideas about the living world. First, that all living things belong to one, and only one, 'natural kind'. An animal is a dog, or a cat, or a badger, and so on: it must belong to some 'species', and cannot belong to two, or to none, and cannot change its species. Second, they share the idea that natural kinds can be classified hierarchically into higher taxa; for example, a dog is a flesh-eater, a mammal (as opposed to a fish, a reptile, etc.), and an animal (not a plant). These universal human attitudes to living things may reflect an innate predisposition. The alternative is that they could be universally believed because they are true, or almost so, and would be learnt by any human society to which a knowledge of the living world was important. A second argument in favour of a special natural history module is the speed with which children acquire these beliefs. But as yet the case for a special module is not decisive.

Thus there is clear evidence for an increase in social, technical, and natural history intelligence in humans, and some hints, particularly from child psychology, of innate and to a degree independent mental modules underlying these skills, analogous to the module responsible for linguistic competence, for which the evidence is much stronger. Mithen points out that, if such modules exist, the extreme conservatism of human prehistory becomes easier to understand. If technical and natural history modules were separate, it would help to explain many features of Lower Palaeolithic tools; for example, the failure to use bone, antler, or ivory in toolmaking, and the uniformity in time and space of stone spear points, despite the range of different food species, in contrast to the great variability of such points manufactured in the past 30 000–40 000 years. Similarly, if technical and social modules were separate, it would explain the total absence of any form of art, or of personal ornament.

The argument, then, is that the increase in human brain size prior to the emergence of modern humans some 100 000 years ago was associated with an increase in social, technical, and natural history skills, but that these abilities were to a large degree independent. Mithen suggests that the competence for language, including grammar, also evolved in this period, although precise dating is obviously difficult. He ascribes the cultural explosion that began some 50 000 years ago, and that has led to continuous and cumulative cultural change, to a breakdown in the isolation between mental modules caused by the emergence of language. The essential point is that, once words exist for social, technical, and living things, the same grammar can be used to say things about them. For example:

The stone hit the nut and caused it to break in half.
John talked to Mary and persuaded her to help him.

In other words, language makes it possible to see the analogies between folk psychology, folk physics, and folk natural history. It is a commonplace that creativity in science and the arts often depends on seeing analogies. If Mithen is right, it was the evolution of language that broke down the barriers between different domains in the mind, and thereby liberated us from the one-million-year conservatism of the Lower Palaeolithic, and made possible the continuous cultural evolution that followed.

Models of human society

Since the time of Plato and Aristotle, philosophers have discussed the nature of society, and how laws can be devised to ensure that it functions harmoniously. It must, therefore, seem somewhat arrogant for two biologists to claim to have something new to say. All the same, we intend to make the attempt. First, we must make it clear what we are *not* going to do. We are not going to argue that human society can be analysed by methods that would be appropriate for an insect society: that is, by explaining the behaviour of individuals in terms of maximizing inclusive fitness (that is, fitness allowing for the effect of an individual's behaviour on the fitness of relatives), and then explaining the structure of the society as a whole as resulting from the interactions of such individuals. The prevalence of cultural inheritance makes such an approach inadequate, although one should not underestimate the role of relatedness in influencing the behaviour of individual humans.

We are, however, going to make use of two ideas derived from biology. The first is that humans, like other animals, have evolved by natural selection, and therefore have predispositions that ought to make sense in terms of past selection. The second is that complex systems can best be understood by making simple models. It may seem natural to think that, to understand a complex system, one must construct a model incorporating everything that one knows about the system. However sensible this procedure may seem, in biology it has repeatedly turned out to be a sterile exercise. There are two snags with it. The first is that one finishes up with a model so complicated that one cannot understand it: the point of a model is to simplify, not to confuse. The second is that if one constructs a sufficiently complex model one can make it do anything one likes by fiddling with the parameters: a model that can predict anything predicts nothing.

We start, therefore, with a simple model of society, which we call the Social Contract game. The essential assumption is that society consists of a group of equal individuals, behaving rationally. If everyone co-operates, each individual is better off than he or she would be if everyone 'defects'—that is, behaves selfishly. The snag is that, in a society of co-operators, it would pay an individual to defect. If everyone else pays their taxes, it would pay an individual not to do so: but if no one pays taxes, there would be no schools, hospitals, or roads, and everyone would be worse off. Hence, to ensure co-operation there must be an element of compulsion. Suppose, then, that everyone agrees to the following contract, 'I will co-operate: I will join in punishing anyone who defects'. Would not this ensure general co-operation, for the benefit of all? Sadly, it would not. The act of 'joining in punishing' would be costly to the individual, even if not greatly so. The society would hence be invaded by 'free-riders', who co-operated, but did not join in punishing.

The contract can be saved by adding a further commitment, 'I will treat as a defector anyone who does not join in punishing'. It seems that the problem of co-operation is solved: as Immanuel Kant once remarked, it is easy to ensure co-operation between even a race of devils, provided only that they are intelligent. Before considering the weaknesses of the model, we must first ask what qualities individuals must have before they could reach such a contract. First, they must have language. Second, they must have a theory of mind: there would be no point in one individual suggesting to others that they agree to the contract unless he assumed that others had a mind like his own, with similar desires, and similar powers of reason.

The Social Contract game suggests that it should not be all that difficult to ensure co-operation between a group of 'equal individuals, behaving rationally'. The weakness, of course, is the assumption of equality and rationality. We return below to the problem of equality, but first we must query the assumption of rationality. As we write, there are many parts of the world where it is obvious to outsiders, and to many of the inhabitants, that almost everyone would be better off if they ceased to identify with subgroups—Muslim, Serb, or Croat; Tutsi or Hutu; Jew or Arab; Protestant or Catholic—and worked together for the common good. Yet a sufficient proportion of the population identify with one or other subgroup, rather than with the human population of the region as a whole, to make such co-operation impossible. Why?

The clue is that group identity, and hence behaviour, is influenced by myth and ritual, as well as, and even to the exclusion of, rational self-interest. Historical myths concerning people's origins, reinforced by ritual, are a powerful influence on human behaviour. Why should this be so? What we are seeking is an evolutionary explanation for a universal human characteristic—the ability to be socialized (or indoctrinated, depending on your point of view) by myth. The particular stories, tunes, apparel, and ritual behaviour that bind a group together are clearly cultural, but the capacity to be influenced by them is innate, and calls for an evolutionary explanation. The obvious one is that human groups that could instil group loyalty into their members would be more successful, and so individuals in the group would transmit more of the genes that made group loyalty possible. But is such an explanation plausible?

Again, as so often in this book, we are faced with a conflict between individual and group selection. Archaeological evidence suggests that early human groups were small, which would make between-group selection more plausible. But if there was extensive genetic exchange between groups, as a comparison with other primates suggests is likely, this would greatly reduce the effectiveness of group selection. It seems likely that the capacity for group identification would not have evolved unless it was selectively advantageous within the group, as well as advantageous to the group in competition with others. How could this be so? A plausible answer is that individuals who did not acquire the norms of the group would be penalized by other group members. There is an obvious analogy between this and the nature of the contract required for stable co-operation in the Social Contract game. In that game, co-operation between rational agents requires punishment of those who defect. In the evolutionary scenario we are now suggesting, co-operation is induced by myth and ritual, not by reason, and individual behaviour depends on an innate capacity to be influenced by ritual. We are arguing that such a capacity will evolve by natural selection if two things are true: groups of individuals with such a capacity are more successful, and, within a group, individuals who lack the capacity are punished, just as defectors are punished in the Social Contract game.

Modern societies consist of millions of individuals, not of tens or hundreds. The monotheistic religions are myths inculcating loyalty to the society as a whole, or to all human beings, not just to a small group, although, as the examples given above show, they can all too easily be distorted to fuel hatred between neighbours. There are subgroups even within societies not divided by language, religion, or history. As we mentioned earlier, a second weakness of the Social Contract game is that it assumes that societies are composed of equal individuals. In agricultural and industrial societies, individuals are not equal. Those who own land, factories, or shops have different options open to them than do peasants, factory workers, or shop assistants. The members of a social class have common interests, and, not unexpectedly, develop myth and ritual to strengthen their struggle to realize those interests: 'The people's flag is deepest red, it's shrouded oft our martyred dead'. Curiously, the cohesion of these groups has weakened during the second half of this century, although the inequalities in wealth on which they are based remain. The likely reason is that today's myth-makers, television and the tabloid press, are not controlled by poorer groups within society. There remain groups like the Freemasons bound together by nothing except ritual and self-interest, and the Mafia, in which kin-bonding is also important. Recently, groups based on sex have acquired a political agenda, and are developing myths to strengthen their cohesion.

As in the other transitions described in this book, the emergence of modern society requires the co-operation of entities that, in the past, were independent and competing. Populations of, at the most, a few hundred individuals, with little division of labour except, probably, that between the sexes, have been replaced by societies of many millions, dependent on extensive division of labour. Co-operation depends both on the rational formulation of laws, or social contracts, in the common interest, and on myth and ritual that instil group loyalty. Unhappily, reason can too readily lead to anti-social self-interest, and group loyalty to irrational xenophobia. We need to create legal systems in which self-interest does not lead to social destruction, and myths that extend loyalty to the human species as a whole.

Miroslav Radman, born in ex-Yugoslavia, has recently written of the hatred and cruelty involved in tribal wars between neighbours. We did not see his essay until this chapter had been written. As we do, he seeks an evolutionary explanation for the human instinct that leads to such wars, and suggests that it may lie in the value of human cultural diversity. Although his explanation is not quite the same as ours, he emphasizes, as we do, the importance of myth and ritual in such conflicts, and argues that we need to develop rituals that generate tolerance rather than hatred. Cultural diversity is greatly to be valued, but we need such rituals if we are not to pay a bitter price for it.

CHAPTER 13

THE ORIGIN OF LANGUAGE

It is impossible to imagine our society without language. The society we live in, day and night, depends on it. Even as we sleep, information about us is being stored, and maybe processed. Imagine that we apply for a job on the other side of the world. We are confident, or at least we hope, that our application will be fairly treated, and that the country to which we would like to move is running properly; that is, that social contracts are observed. Our lives depend on the social division of labour, and on detailed social contracts, which could not exist without language. No ape or dolphin could comprehend, even in spoken form, a contract for a job.

During the past two decades, the conviction has grown that language has a strong genetic component. In some sense, our language capacity must be innate; we can talk and apes cannot, and the reason is that we are genetically different. Yet there are two ways in which language might be innate. We may be just generally more intelligent than apes, and the ability to talk is just a by-product of this fact. Or, and this seems more and more likely, there is a specific 'language organ' in our brain, analogous to a 'language chip' in a computer; this organ is to some degree hard-wired, in that some of its neural connections are set correctly without external stimuli.

It is Noam Chomsky and his school who have contributed most to the lin-

evolution, which has led to three major insights:

natural language—such as Hungarian or English—there is a finite applying these rules, one can generate all possible grammatical language. This list of rules is called a generative grammar. If the of their genetic origin, are able to learn any human it evolved. Three this holds for adults as well, provided they have the main difficulty tongue. Thus there is a general ability to cope rammar, which is called 'universal grammar'.

ate component.

it is natural for biologists to ask how question are formidable. Perhaps excitement, is the uniqueness of our language capacity in the living world. Development is relatively easy to study, because *Drosophila* and mice also develop, whereas not even apes have language in our sense. Our immediate predecessors, such as *Homo erectus* and Neanderthals, are extinct, and language does not fossilize. Things would be easier if our ancestors had started to write as soon as they could talk, but writing is a late invention. In his recent book *The language instinct* (1994), Steven Pinker aptly likened the evolution of language to that of the elephant's trunk—a complex adaptation, unique to elephants, which does not fossilize; yet few scientists doubt that the trunk evolved by natural selection. With language, many people still have doubts, though they are unable to suggest any sensible alternative.

The human condition: brain mechanisms

It has been known for a long time that injuries to particular regions of the brain may cause specific impairments to language (Fig. 13.1). Patients with severe language disorders are called aphasics: the two main types are Wernicke's and Broca's aphasics. The latter have severe difficulties with grammar, whereas Wernicke's aphasics, sometimes called fluent aphasics, produce grammatical sentences with little meaning.

Some of the phenomena associated with injuries of the brain are not only dis-

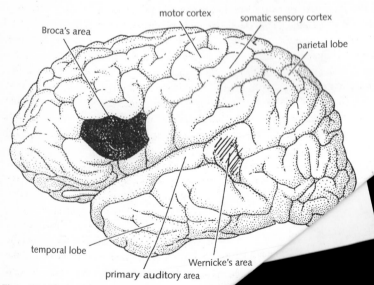

Figure 13.1 The location of the Broca and Wernicke important in linguistic processing.

tressing, but at first sight paradoxical. For example, it is entirely possible for a patient to complain: 'I see the object, I know what it is, but I cannot name it'. It turns out that the formation of a concept, retrieval of the word associated with it, and establishing a link between word and concept, are localized in distinct parts of the brain. Further localizations have been discovered. Some regions of the brain seem to store verbs, other regions store nouns, and there even seems to be a neurological difference between handling nouns referring to animate and inanimate objects.

These observations are consistent with the innateness of the language organ, but they do not prove it. Imagine one starts with a network of interconnected neurones, capable of 'learning' (by strengthening the connection between two neurones if they are active simultaneously, and weakening the connections if they are not), but without any localization. It has been shown by computer simulation that, as such a network learns, some parts will, by sheer accident, become associated with particular tasks. If one adds to this model the fact that particular sensory inputs (hearing, seeing, etc.) are localized, as are particular outputs (for example, speaking), then not only will functions within the net become localized, but the same localizations will appear in different nets, or, by analogy, in the brains of different individuals.

Thus localization in the brain suggests, but does not prove, innateness. There is a second, related problem: how does the brain learn to follow grammatical rules? How, for example, do we store the rule for forming the past tense of a verb (add -*ed*)? There are two extreme possibilities. One is that rules govern everything: not only the typical case, verb + *ed*, but also *ey* → *oo* as in *take, shake, forsake,* and *i* → *a* as in *sing, ring, spring*. The alternative is that every case must be learnt as a separate item, just as the meanings of words must be learnt. Pinker (among others) suggested a fruitful reconciliation of these two extreme views. There may be a rule for the regular verbs, whereas irregular verbs are individually memorized, as 'dictionary items'. This would explain how verbs with the same stem can have different past tense forms; for example, *ring/rang* the bell versus *ring/ringed* the city.

The idea that separate brain processes are involved in forming the past tense of regular and irregular verbs is supported by the study of patients with neurological disorders. The work depends on the notion of 'priming'. It is found that the speed with which a person can recognize the word *goose* is greater if they are first primed with the word *swan*. This is what one would expect if the words are stored close to one another, or, more generally, if there is some link between the storage of one and of the other. On both the 'everything is rules' view, and the 'everything is learnt' view, we would expect that *walked* should prime *walk*, and that *found* should prime *find*: after all, there is a connection between the members of a pair, whether it is rule-governed or learnt. But W. Marslen Wilson, and

Loraine Tyler observed patients in whom *found* primed *find*, and *swan* primed *goose*, but *walked* failed to prime *walk*. In another patient, they found the exact opposite: *walked* primed *walk* but *swan* failed to prime *goose* and *found* failed to prime *find*. These patients had damage in different brain areas. This correlates well with the earlier finding that patients with agrammatic aphasia (who find composing sentences difficult) have more trouble inflecting novel regular verbs than irregulars, whereas anomic aphasiacs (who find retrieving words difficult) have more trouble with irregular verbs.

What all this suggests is that the ability to form the past tense of irregular verbs depends on the same mechanism as learning the meanings of words (each case must be learnt as a dictionary item), and is different from the ability to form the past tense of regular verbs (a rule has been learnt).This by itself says little about innateness, but at least reveals that language rules do have a manifestation in specific areas of the brain.

We have spent a lot of time on what may seem a trivial question—how do we form the past tense of verbs? But the research does show how a combination of neurological and psychological studies is beginning to reveal how the language organ works, and also how hard it is to discover in what sense it is innate.

Language acquisition

A key question is whether we learn language simply by 'trial and error', as the behaviourists would have liked us to believe, or whether there is some instinctive 'knowledge' hard-wired into the human brain, which has a genetic basis. Interestingly, the latter idea is not entirely new: it emerged in the seventeenth century in 'philosophical grammar', but was forgotten after the rise of Romanticism. Clearly, language does not develop in full deprivation of 'linguistic input': if nobody talks to you, you won't learn to talk. There have been some remarkable cases in history suggesting that there is a 'critical period', ending about puberty, when the window for learning a mother tongue closes. If you miss your chance to learn it during the critical period, you will never be able to master any natural language properly. A sad but well-documented case is that of Genie, a girl whose father kept her locked away for many years. Although later she was taught to speak, she never got beyond the level of broken, rather ungrammatical English. So linguistic input seems an absolute necessity. But input necessary for normal development is not a valid argument against a considerable degree of innatism, as can be illustrated by analogy with the development of vision.

The visual system—the eye, the optic nerve, and the associated brain regions—is an extremely sophisticated complex of organs. No one doubts that it is an evolved adaptation, increasing the fitness of the organism equipped with it. It is largely innate: mutations are known affecting different components, from colour vision to anatomical details. First, by analogy with language, we can ask two questions. What happens to the visual system if there is no input early in development? Second, if all its components are needed if it is to work properly (a lens is no use without a retina), how could it ever have evolved? The answers are now known and pleasing; we discuss them in turn.

Young cats, when blindfolded, do not develop proper vision. If one removes the cover from the kitten's eye before 8 weeks, normal development resumes, but if this happens later, impairment is guaranteed for life. Thus there is a critical period during which the visual system must receive external stimuli for its development. The neuroanatomy of this system is comparatively well known. It turns out that synaptic connections between neurones develop in response to stimuli, and are eliminated in their absence. Thus in cats there is an innate vision acquisition device (VAD), which, given appropriate visual input, produces proper vision, useful for the organism. The capacity to develop vision requires a specifically pre-wired part of the brain: other parts cannot serve as substitutes. Presumably, setting up a proper visual system is such a difficult task that it is more readily achieved if the stimuli themselves are used as cues in the process. By analogy, acquisition of language requires linguistic input, but it may also depend on pre-wired parts of the brain. A system, visual or linguistic, can depend on suitable input early in life, and also depend on genetically determined brain structures.

Turning to the second question, the evolutionary origin of the eye has fascinated generations of biologists and non-biologists. William Paley used it in his natural theology as part of the argument from design. Whereas biologists accept that there is a lot of good 'engineering' in the eye, the engineer was, to borrow Richard Dawkins's successful metaphor, surely blind: it was evolution by natural selection of heritable variations. A dramatic demonstration of this principle was given by Dan Nilsson and Susan Pelger. They have shown that a smooth evolutionary pathway spans the space between a simple group of light-sensitive cells, as found today in some simple invertebrates, and a 'perfect' camera eye, similar to those of humans and cephalopods. The whole 'journey' through the sequence of intermediate structures can be taken (assuming moderate heritability, weak selection, and small genetic variation for the traits in question) in about 400 000 generations. A rudimentary eye, even if remote from our marvellous organ (or that of an octopus, which, curiously enough, looks very similar) in its complexity, is much better, from an adaptationist point of view, than none at all, if the animal needs to see. It is especially lucky that eyes of intermediate levels of complexity still exist in animals: reconstruction of the evolutionary history of the eye is an easy task compared with that of the 'language organ'. Language not only does not fossilize, but there are no living intermediate forms either. Evolutionists insist, however, that such intermediates must have existed.

Thus three conclusions emerge from the eye story: (1) it is easier to inherit a 'vision acquisition device' than a full-blown hard-wired visual analyser; (2) the visual analyser, once 'set up', is refractory to radical restructuring—hence the existence of a critical period in its development in cats; (3) the eye seems to have evolved in steps from a light-sensitive, innervated cell to our complex organ by common evolutionary mechanisms.

Something similar may have been taking place in evolution of the language organ, and may be occurring during individual development. An argument, put forward forcefully by Noam Chomsky and his followers, refers to the 'poverty of stimulus'. Most permutations of word order and grammatical items in a sentence leads to incomprehensible gibberish. There is no way that children could learn without some internal 'guide' which sentence is grammatical and which is not, only on the basis of heard examples. To make matters worse, many parents do not correct their children's grammatical mistakes (they seem to be much more worried about the utterance of four-letter words). Recent investigations clearly confirm that children's 'instinctive' understanding of grammatical intricacies, between the ages 2 and 4, is far better than one would expect from a conventional learning mechanism. Thus there seems to be a 'language acquisition device' (LAD) in the brain, which must be triggered by linguistic input so that its working ultimately leads to proper language. It is the LAD, and *not a fully developed linguistic processor*, which seems to be innate.

Tuning the language organ

So far so good. But why is it that children have to learn their language for many years, and why, despite admirable progress, do they continue to make grammatical mistakes during this period? Why are people without 'linguistic input' unable to master language properly as adults? Why do we have to learn grammar at all—why is it not all hard-wired? Why are there so many languages, not only in terms of vocabulary but also in terms of their generative grammar? Why is universal grammar insufficient by itself? We will attempt answers to these questions, and also to the question of whether the language instinct could have arisen by the conventional evolutionary processes of mutation and selection.

Perhaps the easiest of these questions to answer is why our vocabulary is learnt rather than innate. If words were innate, cultural evolution could not be faster than genetic evolution: linguistic innovations, such as the word *screwdriver*, would have to be genetically assimilated before they could be used. We need a grammar whereby any type of statement can be made, but the actual contents of each statement should be cultural rather than genetic.

The reason why there are many different natural languages, which have to be learnt in a critical period of development, seems to depend on two considerations: language must be learnt as soon as possible, and learning a language is easiest, and perhaps only possible at all, using the LAD (language acquisition device) method. The maintenance of the latter is presumably costly in time, because it requires that synapses be left plastic and options be left open.

Different generative grammars—for example, Hungarian and English grammar—are alternative modes of operation of the same language organ. This raises the question, are some grammars easier to learn than others, because the language organ develops some neuronal configurations easier than others? We do not know, but the work of Derek Bickerton on pidgin and creole languages suggests that the answer may be 'yes'. Pidgin is a means of communication that emerges when adults with no common language come in contact, as happened frequently in communities of 'wage slavery' in various former colonies.

We discuss pidgin in more detail below; for the present it is sufficient to say that it is a limited means of communication without grammar. Creole languages emerge when children grow up in such communities, where the main linguistic input is pidgin. Such Creole languages are proper languages, with a fully developed grammar. The startling claim is that the emerging grammars, although there is variation, are very similar in different communities, even if they are several thousand kilometres apart; one example is that the 'double negative' is usually grammatical in Creoles. One could still argue that in these cases the parents did speak proper languages, and that their influence was decisive in the emergence of the new Creole grammar. Since then, however, an independent line of evidence has shown that groups of children can evolve a language with a proper grammar, even when the only outside linguistic input is pidgin-like. Groups of deaf children learning sign languages under the influence of parents who can sign only at a pidgin level go far beyond the grammatical level of their parents, developing a sign language with a sophisticated grammar, even if there is no adult proficient in sign language to talk to. Creole languages are illuminating in two respects. They illustrate the linguistic creativity of communities of children, and, by their common features, they provide hints as to what particular generative grammar emerges most readily, given our common universal grammar.

What is really innate, or the nature of universal grammar

A widely accepted idea is that, in every language, sentences are built by combining noun and verb phrases in an appropriate way. This is explained in more detail in Fig. 13.2; English readers can get a more immediate appreciation of the idea by reciting the nursery rhyme, 'This is the house that Jack built'. The

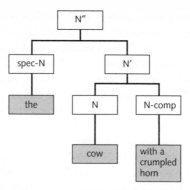

Figure 13.2 The structure of a noun phrase. N (*cow*) is the head of the phrase, and must be a single word. It is first linked to its complement (*with a crumpled horn*) via the node N', and N' is then linked to the specifier (*the*) via the node N'', representing the full phrase. Although the head, N, must be a single word, the complement may itself be a phrase, as it is in this example. Thus phrases are like Chinese boxes, stacked inside one another.

actual rules in the generative grammar for a particular language can sound rather complicated. To give but one example: we know that the question

How do you know who he saw?

is grammatically correct, whereas the question

Who do you know how he saw?

is incorrect. Although most people with a moderate command of English can see this, not even highly literate native English speakers could explain why this is so, unless they happen to be linguists. The explanation is given in terms of null elements. The second question should be written *Who do you know how he saw —?*, where — marks the place from which the object of *saw,* now replaced by *who,* has been moved. But, according to linguists, there is a constraint on movement, asserting that a *wh-* word cannot be moved across a space occupied by *how.*

This constraint is subtle one, but linguists argue that the assumption of null elements is the simplest way to explain this and many related phenomena of grammaticality. They may be right, but we cannot be sure that there is no simpler rule to account for the facts. Or perhaps evolution chose a more complicated system for historical reasons; if so, the rule we are really following would be more complex than that suggested by linguists. In any case, it is intriguing that, when talking, we obey without effort a rule of which we are not conscious.

A new initiative in linguistics, favoured by Chomsky himself, is the so-called minimalist programme, which has the aim of formulating the simplest possible rules specifying how words should be combined to form grammatical sentences. We cannot go into details, but, following Robert Berwick, we will try to give a flavour of this approach by drawing an analogy with the laws of chemistry. This may seem an odd thing to do: after all, most people know less of chemistry than they do of language. Yet the laws of chemistry are rather well understood, whereas the nature of universal grammar is still largely a mystery.

First, there is a reason why we might expect such an analogy to help. In chemistry, a small number of kinds of atoms—hydrogen, oxygen, iron, and so on—can combine to form an immense number of different compounds, yet by no means all arrangements of atoms form stable compounds. In the same way, a small number of kinds of words—nouns, verbs, adverbs, and so on—can combine to form an immense number of sentences, yet not all combinations are grammatical. So there may be an analogy, and, as any reader of this book will have discovered by now, we are great believers in the power of analogies to generate useful ideas.

Chemistry is a powerful and sophisticated branch of science, largely opaque, as it happens, to one of the authors of this book. Yet the rules governing how atoms combine can be simply explained. The first is the concept of valence. Each kind of atom can be visualized as having a number of attachment points, from one to four, by which it can be linked to other atoms to form a molecule. The number of these attachments is known as its valence. For example, hydrogen (H) has valence 1 and oxygen (O) has valence 2. Hence one can form a molecule of water by linking one oxygen to two hydrogen atoms, to give H–O–H, or H_2O, but one could not link two oxygens to one hydrogen, to give O–H–O. Of course, there is more to it than this, but it remains true that simple rules of combination are obeyed. A second rule one might call the 'subassembly' rule. Once a molecule has been formed, it can behave as a unit, linking to other molecules according to rules similar to those that govern the linking of atoms to form molecules. For example, once nucleotides have been formed, they can be linked end to end to form a more complex molecule, DNA: when this is done, precisely one molecule of water is removed from each link.

The possible analogy with language should now be clear. There are rules that govern how words can be linked to form sentences. Further, phrase-structure grammar (Fig. 13.2) implies that a single word can be replaced in a sentence by a group of words: thus *cow* can be replaced by *cow with a crumpled horn*. The idea, then, is that words have a property analogous to valence, enabling them to combine only with words that match. There is an operation, *merge*, analogous to chemical reaction, that combines words into sentences. If *merge* is impaired, one has ungrammatical aphasia. If there is a problem with word features, analogous to valences, we have feature-blind grammar, which is described in the next section.

In this analogy, there is a difference between language and chemistry in the way in which meaning emerges. There is a difference in the meanings of *The dog bit the postman* and *The horse ate the grass*, although they have identical grammatical structure. In any language there are many nouns, whose meanings have to be learnt, but the rules governing their combination—their valencies—are the same.

The short answer to the question at the head of this section is that we do not know the nature of universal grammar. We cannot discover it simply by introspection. Will we ever know the answer? As biologists, we suspect that, in the long run, the most powerful tool for discovering the answer may be genetics.

Genetics

If our language faculty has an innate component, then there should be genetic variation for this trait. As we all know from our school years, there is a marked quantitative variation of linguistic skills in the human population, both in our mother tongue as well as in foreign languages. How much of this observed variation is genetic is not known, and would be hard to test. Qualitative variation, leading to some well-defined impairment of grammar, should more readily analysable. For this to be so, however, the impairment should be specific to language. It is easy to imagine that linguistic impairment would be associated with, and partly caused by, other deficiencies, such as a diminished IQ or deafness. We need to find cases in which these associated impairments occur without the language deficit; and, conversely, the language deficit should not always be accompanied by these complications. This is called a double dissociation.

Although such variation must have been around for tens of thousands of years, the first clear evidence of genetic involvement was published by Myrna Gopnik only a few years ago. The case concerned an English-speaking family exhibiting a strange type of language problem (dysphasia): they have problems with grammatical features such as past tense and plurals. Let us quote a few sample sentences:

She remembered when she hurts herself the other day.
Carol is cry in the church.
On Saturday I went to nanny house with nanny and Carol.

The problem is obvious: regular grammatical features, such as *-ing*, are missing. Note that the irregular verb, *went*, is used correctly, because the affected individual had to, and could, learn it as a special case, just like everyone else. Some members of the family have trouble in completing simple tests such as the following. An individual is shown a picture of a strange animal, and told that it is a *wug;* shown a picture of several such animals, and asked what they are, the individual cannot find the answer, *wugs.*

It seems that what the affected individuals cannot do is to learn general rules about how the form of a word must be changed to express present/past, or singular/plural, or possession (*nannie's* rather than *nanny*). Thus they can learn as a dictionary item that, for example, the past of *watch* is *watched*, but cannot generalize to other verbs; they do not then know that the past of *wash* is *washed*. The family pedigree (Fig. 13.3) shows that this impairment is caused by a single dominant gene form (allele) on an ordinary rather than on a sex chromosome. Since then, French, Japanese, and Greek cases have also been found, although more genetical work would be welcome.

This interpretation of the study has not been universally accepted.The most frequent objections are that the patients have a problem with auditory and/ or articulatory processing and that grammatical complications arise as a by-product; or that we are faced with a general cognitive problem. We discuss these in turn.

It could be, for example, that patients, due to a failure in auditory processing, cannot hear the small /-d/ sound at the end of the verbs in past tense. Yet tense is marked differently in the other languages, and the problem remains. Furthermore, the patients make the same mistakes in writing as well. When they say something like *walked,* this is not an automatic *walk* + *-ed* for them, but a separate word, or a *consciously* produced form. For example, after lengthy speech therapy, these patients can learn that for plurals they have to add an /-s/ sound to the word (that is, a 'hiss', as in *plates* or *claps*), but apparently they do it via a route different from ours: they form the plurals of 'sas' and 'wug' so that they sound 'sasss' and 'wug-s', respectively, whereas we would say 'sassez' and 'wug-z'.

As to the idea that there is a general cognitive problem, it is true that there are patients who have other problems, but most of them do not; for example, one linguistically impaired individual is a rocket scientist. Just as people who are linguistically impaired often have a normal IQ, it is also true that most people with low IQ, or with other cognitive difficulties, have an unimpaired language faculty. Further, there are neurological data indicating that the linguistically impaired have an anatomically identifiable distortion in the brain. It is likely that it can cause the language impairment, either alone or in association with other problems; even if the second possibility is true, it does not follow that these associated problems are the cause of the linguistic impairment, which also occurs in their absence.

The past two decades have witnessed a remarkable advance in the genetics of development (see Chapter 10). Now we are able to dissect genetically various component processes by developmental mutants: it is the malformations that

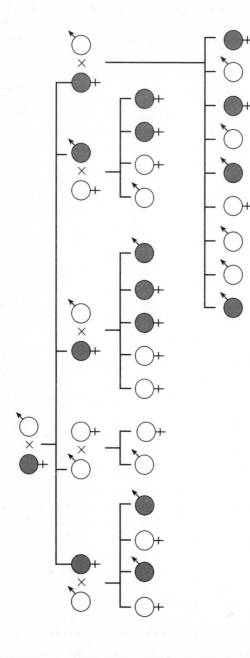

Figure 13.3 Pedigree of a family with a specific language impairment, after Myrna Gopnik. Affected individuals are shown by filled circles; ♂ indicates a male, and ♀ a female.

reveal best how the normal system works. We expect similar progress towards a science of linguistic genetics. The job is much harder, however, because language is unique to humans, so that, instead of being able to cross individuals at will, or elicit mutations in them, we are restricted to the analysis of existing family pedigrees.

From alarm calls to the Tower of Babel: evolution of the language capacity

The evolutionary gap to be bridged is between 'protolanguage' and Chomskyan universal grammar. Examples of protolanguage include pidgin, the language of children under two, the learnt language of apes, and the language of people brought up in linguistic deprivation. Some examples are given in Table 13.1. Its characteristics include the following:

1. The use of words as Saussurean signs (Fig. 13.4); that is, a word must stand for a concept, both for the speaker and for the hearer.

2. The lack of purely grammatical items, such as *if, that, the, when, in, not* that do not refer to anything.

3. The absence of hierarchical syntax—for example, the use of phrases discussed above.

Although lacking these essential features of language, such protolanguage has already a lot. In particular, most linguists would argue that there is no convincing example of proper words used by animals *in the wild*. Consider, for example, the signals that vervet monkeys have for martial eagles, leopards, and pythons. It is accepted that adult monkeys give different calls when they see these different predators. Juvenile monkeys know without learning that the eagle call, for example, should be given in response to flying objects, but must learn its precise application; at first, they may give the call to other, harmless birds, and even to a falling leaf. Monkeys hearing such a call respond appropriately; it is a good

Table 13.1 Examples of protrolanguage: A, utterances of 2-year-old children; B, signed statements of a chimpanzee raised by humans; C, utterances of Genie, a girl deprived of linguistic input until the age of 13 years

A	B	C
big train	drink red	want milk
red book	comb black	Mike pain
Mommy lunch	tickle Washoe	At school wash face
go store	open blanket	I want Curtis play piano

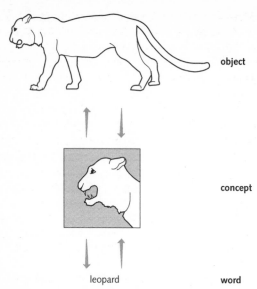

object

concept

leopard · · · · · · · word

Figure 13.4 The structure of a Saussurean sign. Arrows indicate a direct connection. Note that there is no direct connection between an object 'out there' and the corresponding word: concepts stand in between. If one either hears or sees a leopard, that elicits in the mind the concept of a leopard, which in turn elicits the word 'leopard': in reverse, if one hears the word 'leopard', this elicits in the mind the concept of a leopard, and one can then imagine the sight or the sound of a leopard. To qualify as a word, a signal must be such a sign: all the arrows must be present. Whereas concept formation is, to varying degrees, quite common in animals, it is unclear how often words are used. For example, if a vervet monkey hears the alarm call for a snake, does it first form the concept of a snake in its head, or does it just respond? We don't know. After all, it would be easy to design a machine that responded in a specific way to a specific sound, without forming anything corresponding to a concept.

idea to climb to the tip of a branch if there is a leopard about, but not if there is an eagle. But, and this is the crucial point, there is no evidence that when a vervet hears the alarm call for python, it actually forms the concept of a snake in its brain; it may respond without knowing why it does so. Thus one arrow may be missing from the diagram in Fig. 13.4.

Other animals seem to be able to use real Saussurean signs, maybe even in the wild. One of the suspects is the dolphin. Remarkably, the size of the brain relative to body size is second only to humans: the great apes have only half the score of dolphins. This by itself may be of limited significance; after all, Neanderthals had a bigger skull than we have. Dolphins, however, have a very rich vocal repertoire that seems to be open-ended: a dolphin can learn new signals until the end of its life. Dolphins can listen to the signals of conspecifics and react appropriately by extracting the relevant information. Some of the signals seem to be 'signatures' identifying the emitter. It is very likely, although not yet proven, that dolphins can use Saussurean signs—or words for short. For example, when a dolphin 'says' something in reaction to a shown object, another dolphin is able to pick another piece of the same object without knowing what the first individual saw.

They also seem to have an understanding of the importance of word order, and hence have been credited with 'sentence comprehension'. For example, they can be taught to distinguish properly between *pipe fetch hoop* and *hoop fetch pipe*. They are able to react to several hundred 'sentences' of two to five 'words' in length and perform appropriate actions. Once they get accustomed to a syntax such as Direct object + Action + Indirect object, they can use this template with novel words at the first instance. For example, having learnt that the sentence *hoop fetch pipe* means 'fetch the hoop to the pipe', and also the meanings of the words *net* and *basket*, it seems that a dolphin will correctly interpret the sentence *net fetch basket* the first time it is experienced as meaning 'fetch the net to the basket'.

This performance is remarkable, but there are three limitations:

1. Comprehension in general is easier than production: as we all know; it is easier to understand than to explain.

2. The dolphins did not produce sentences in the experiment.

3. We do not have the faintest idea whether they use words, let alone syntax, in the wild, whatever they may do in captivity.

Mária Újhelyi called attention to the fact that the songs used in territorial defence by some monkeys that live as monogamous pairs, and by gibbons, which have a similar social system, qualify as a pre-linguistic systems, in the following sense. They combine discrete elements in different sequences, and these sequential differences have a meaning, such as the signing of sex, identity, territory, and so on. The songs are particularly fascinating in gibbons, where a phenomenon called duetting also occurs between members of a 'married couple'. It seems that chimpanzees and bonobos (pigmy chimpanzees) retained this pre-linguistic faculty, as exemplified by their long calls, although they are not monogamous. An outstanding question is, of course, the referential nature of such vocalizations: what do they actually mean?

As must be clear by now, we think that there must have been intermediate stages of grammar. It is not hard to think of intermediates. For example, David Premack has suggested that there could have been a stage when it was be possible to say *the dog bit John*, but not that *John was bitten by the dog*. That is, the subject of the sentence had to appear first, and had to refer to the active 'agent'.

In fact, one can suggest a number of steps by which language could have been extended:

- Items for negation, such as *no*.
- '*Wh-*' questions, such as *what*, *who* and *where*.
- Pronouns (instead of repeating proper names).
- Verbal auxiliaries such as *can* and *must*.
- Expression for temporal order (e.g. *before* and *after*).
- Quantifiers such as *many* and *few*.

Of course, without any one of these additions, there would have been things one could not say. But so what? As we have seen, an imperfect eye can be a lot better than no eye at all.

Yet some linguists, including Derek Bickerton, argue that much of syntax must have originated abruptly in evolutionary terms. Novelties in evolution can appear rather suddenly if they are modifications of a structure that evolved to perform some other function. For example, we suggested in Chapter 4 that the genetic code may have evolved from a system in which amino acids functioned as coenzymes of RNA enzymes. Similarly, the feathers of birds first evolved for thermoregulation rather than flight. But when this happens, the old structure will at first be rather inefficient in its new role, and will require fine-tuning by selection.

Bickerton has suggested that syntax may have evolved by the connection of two pre-existing faculties, one being a social 'cheater detector', and the other a protolinguistic ability. This would be a kind of symbiosis, but between two genetic systems in the same organism, not between previously independent organisms. We have already argued that symbiosis can be the source of sudden novelty. If this idea, or something like it, turns out to be correct, it would help to explain both the suddenness and the complexity of human language. A crucial part of the idea is that the 'Machiavellian' thinking of which non-human primates seem to be capable must have had a syntax with some equivalent of phrase structure. This sounds credible: it is difficult to imagine a good Machiavellian who could not think about who did/does/will do what to whom and why, and who could not do this recursively; for example, if I tell Joe, then Joe will tell Mary.

Even if this explanation is correct, there would still have been a need for substantial evolutionary fine-tuning. Think again of the example of organelle evolution: without the protracted phase of evolution leading to specific transport systems, metabolic utilization of photosynthates and ATP would have been impossible, even if a lot was given free by symbiosis.

This idea of a connection between the protolinguistic and the social modules in the brain is similar to the ideas of Steven Mithen described in Chapter 12. According to him, the early human mind, around 100 000 years ago, consisted of modules of domain-specific mentality: social intelligence, intelligence for natural history (hunting, food gathering, etc.), technical intelligence, and language (the latter linked to social intelligence). Archaeological data show that, some 50 000 years ago, there was a great spurt in technical inventiveness and artistic creativity. Mithen suggests in his book *The prehistory of the mind* that the cause of this spurt was an increase in communication between the previously rather isolated modules: for example, the striking increase in the range of tools made at that time required that people think simultaneously about hunting and toolmaking. Language would have helped such communication; there is a sense in which thinking is talking to oneself.

The charm of these ideas is that they are in principle testable by looking at the function, development, and breakdown of specific parts of the brain. It would be particularly persuasive if some neurological disorders turned out to arise from the decoupling of these modules.

One puzzle remains: how could grammatical novelties spread in a population? There is no point in one individual using a new phrase or construction if others do not understand. Would not the novelty be selected against as a 'Hopeful Monster': hopeful in words but hopeless in reality? The first thing to note here is that when we meet a linguistic novelty, we do not give up too easily: we try to guess the meaning by watching others, as well as by trying it out ourselves. Second, grammatical novelties must be built on pre-existing neuronal structures, and so are likely to be compatible with what is already present, just as the latest (and usually more elaborate) forms of computer software tend to be compatible with previous editions.

So a new mutant extending linguistic competence would survive, because others would learn and adopt the novelty. But if the mutation is to spread by natural selection, it must confer an advantage. Why should the new mutant actually be fitter, if others can do by learning what the mutant is hard-wired to do? Two linguists, Steven Pinker and Paul Bloom, have suggested that a possible way out of this trap is by 'genetic assimilation learning'. This is an evolutionary process that can convert a behaviour that is learnt into one that is genetically programmed, without supposing that acquired characters are inherited. It can best be explained by describing a fascinating computer simulation by G. E. Hinton and S. J. Nowlan. They suppose that, in order to perform some action, neuronal switches must be correctly set. The switches can be set by genetics or by learning. But if we depend on genetics alone, a population will never evolve the capacity to perform the action, because the chance that a genotype specifying all the correct settings will arise by random mutation is vanishingly small. Even if such a genotype does arise, it will be broken up by genetic recombination in the next generation. With pure genetics, having 99 per cent of the switches set correctly is no better than having only 10 per cent in the correct mode.

The situation changes dramatically if we assume that learning can occur during the lifetime of an individual; that is, only some of the switches are set by genetics, and many random trials can be done on the switches that are not so set. The earlier in its lifetime an individual hits on the right combination, the more offspring it will send into the next generation. Thus, if a higher proportion of switches are genetically correct, the expected time until the rest of the correct settings are found decreases, so that the expected number of offspring increases. When learning is combined with genetics, having 90 per cent of the switches genetically set in the correct mode is a lot better than having 70 per cent of them set correctly. Thus, a trait that is adaptive, and that *must be learnt* in the first place, can evolve to be hard-wired into the brain, because learning can guide natural selection.

The suggestion, then, is that genetic assimilation can help to explain the evolution of our hard-wired competence to acquire grammar. New components of grammar would first be tried out by individuals, just as new turns of phrase are tried out today, and learnt by other members of the population. If skill in communication increases fitness, then those who learnt new grammatical tricks fastest would leave most descendants, and the initially learnt grammatical novelties would be genetically assimilated.

Are there any other human faculties whose evolution may have positively influenced that of language? A serious suggestion is that the skill to manipulate objects and combine them so that the outcome 'makes sense', so characteristic of humans, may have coevolved with the language faculty. The point is that in purposeful object manipulation an 'action grammar' is apparent. For example, take the sentence *(I) want more grape juice*. The grammar follows what can be called a subassembly strategy: *more* combines with *grape juice* to form a noun phrase, and then the latter is combined with the verb. A similar strategy is apparent, for example, in eating with a spoon: one has to combine the food with the spoon, and the combination has to be put into the mouth.

The analogy goes deeper than this. The neuropsychologist Susan Greenfield has observed children playing the familiar game in which a set of cups of graded size must be fitted one inside another. She finds that action grammar develops in stages that resemble the acquisition of language grammar. Children adopt three strategies for the former: the pairing, the pot, and the subassembly method (Fig. 13.5). These strategies are arranged in an order of increasing complexity that agrees with the temporal order of their emergence in development. The analogous linguistic phases of sentence construction come later in child development. Greenfield points out, however, that analogous stages

Strategy 1
Pairing method

Strategy 2
Pot method

step 1

step 2

Strategy 3
Subassembly method

step 1

step 2

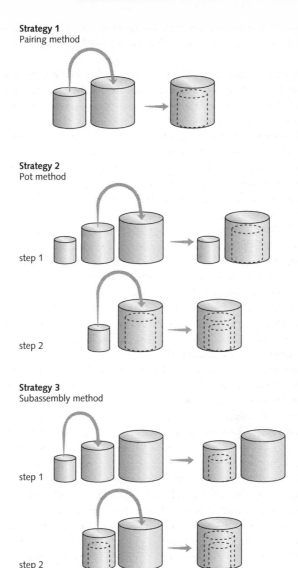

Figure 13.5 Action grammar. Three ways in which a child can arrange a set of cups, with small ones inside larger ones. These have formal similarities to ways of constructing sentences. The pairing method is similar to a simple sentence such as *Mary ate the fish*. In the pot method, two sentences are joined, as in *John caught the fish and Mary ate the fish*. Finally, in the subassembly method, a phrase is first formed, and then used as part of a sentence, as in *Mary ate the fish which John had caught*.

occur as a child acquires the ability to form words by combining phonemes, and suggests that the real synchrony is between action grammar and word construction.

It is remarkable that apes do not use the subassembly strategy for object manipulation in the wild, although some of them seem to have mastered it in captivity. It is of significance that two chimpanzees that discovered the subassembly strategy in the nesting-cup experiments had been exposed to intensive linguistic training. It is perhaps even more important that the only demonstrated example of teaching (not to be confused with learning) in the wild happens in chimpanzees when the mother teaches her offspring how to crack a nut with a hammer and an anvil. Thus Greenfield imagines a coevolutionary process between teaching, using more and more words and grammar; and using tools to perform more and more complicated tasks. This is plausible. One should not forget, however, that one important aspect of language is that we can talk about things that we could never do. To perform complex meaningful actions we must go through many impractical ones in our head: to do good solid science we need well-developed fantasy and imagination.

Language and the future

As we approach the end of our story, we want to reflect on what happened in evolution, and on what may happen next. In several of the major evolutionary transitions one can see either that a novel type of inheritance system arises, or that a system with limited heredity, able to encode only a small number of alternative messages, evolves into one with potentially unlimited heredity. Different inheritance systems include:

- Autocatalytic cycles and networks, as described in Chapter 1 and illustrated in Fig. 1.2.

- Small oligonucleotides; that is, strings of only a small number of nucleotides.

- RNA-like and DNA-like molecules, consisting of some hundreds of nucleotides.

- Chromosomes, like those of bacteria, with only a single origin of replication.

- Chromosomes with many origins of replication, as found in eukaryotes.

- Primitive states of gene regulation, such as the *lac* operon described in Chapter 10, in which the regulated state—on or off—is copied when the cell divides.

- Advanced gene-regulation systems, found in multicellular organisms (for example, it has been suggested that a gene expressed in one tissue type may have regulators for stage of development, tissue specificity, cell-lineage identification, cell-cycle tuning, amplitude control, and reception of signals from adjacent cells).

- Protolanguage.
- Language.

The analogy between the genetic code and human language is remarkable. Spoken utterances are composed of a sequence of a rather small number of unit sounds, or phonemes (represented, at least roughly, by the letters of the alphabet). The sequence of these phonemes first specifies different words, and then, through syntax, the meanings of sentences. By this system, the sequence of a small number of kinds of unit can convey an indefinitely large number of meanings. The genetic message is composed of a linear sequence of only four kinds of unit. This sequence is first translated, via the code, into a sequence of 20 kinds of amino acid. These strings of amino acids fold to form three-dimensional functional proteins. Through gene regulation, the right proteins are made at the right times and places, and an indefinite number of morphologies can be specified.

Thus in both systems a linear sequence of a small number of kinds of unit can specify an indefinitely large number of outcomes. But there is one respect in which the two systems cannot usefully be compared. In language, the meanings of sentences depend on the rules of syntax. These rules are formal and logical. In contrast, the 'meaning' of the genetic message cannot be derived by logical reasoning. Thus, although the amino acid sequence of the proteins can be simply derived from the genetic message, the way they fold up to form three-dimensional structures, and the chemical reactions that they catalyse, depend on complex dynamic processes determined by the laws of physics and chemistry. It does not seem possible to draw a useful comparison between the way in which meaning emerges from syntax, and that in which chemical properties emerge from the genetic code.

Are there ways in which a system of unlimited heredity could work, other than by a linear sequence of a small number of kinds of discrete units? There seems to be no necessary reason why the message should be one-dimensional, except that a linear sequence is easy to arrange, and it is sufficient. But the discrete, digital nature of the units is probably necessary. If meaning was conveyed by signs that could vary continuously, instead of belonging to one of a small number of classes, meaning would be gradually lost, as in the game of Chinese whispers. However, it seems that human language does not depend on phonemes, or their equivalent. Thus the sign languages invented by the deaf do not involve a one-to-one correspondence between signs and phonemes, although 'words' do exist. It would be interesting to know how far these languages are digital.

We have treated the origin of language as the last of the major transitions. This shows that we are biologists, not historians. Language was indeed the last transition that required biological evolution, in the sense of a change in the genetic message. But there have been two major changes in the way in which information is transmitted since the origin of language. The first was the invention of writing. Without writing, or some equivalent way of storing information, large-scale civilization was impossible, if only because one cannot tax people without some form of permanent record. The latest transition, through which we are living today, is the use of electronic means for storing and transmitting information. We think that the effects of this will be as profound as were those after the origin of the genetic code, or of language, but we are not rash enough to predict what they will be. Will our descendants live most of their lives in a virtual reality? Will some form of symbiosis between genetic and electronic storage evolve? Will electronic devices acquire means of self-replication, and evolve to replace the primitive life forms that gave them birth? We do not know.

FURTHER READING

The books listed below discuss some topics—for example, molecular biology, development, language—in greater detail. Perhaps more important, their authors are in a real sense our intellectual ancestors: their books contain ideas which we have tried to develop further.

Crick, Francis (1981). *Life itself: its origin and nature*. Simon & Schuster, New York.

Dawkins, Richard (1989). *The selfish gene*. Oxford University Press, Oxford (first edition, 1976).

de Duve, Christian (1995). *Vital dust: life as a cosmic imperative*. Basic Books, New York.

Dyson, Freeman (1985). *Origins of life*. Cambridge University Press, Cambridge.

Gánti, Tibor (1987). *The principle of life*. Omikk, Budapest (first edition, 1971, in Hungarian).

Jackendoff, Ray (1994). *Patterns in the mind: language and human nature*. Basic Books, New York.

Maynard Smith, John (1986). *The problems of biology*. Oxford University Press, Oxford.

Monod, Jacques (1997) *Chance and necessity*. Penguin Books, London (first edition, 1970, in French).

Pinker, Steven (1994). *The language instinct*. William Morrow, New York.

Schrödinger, Erwin (1993). *What is life?* Cambridge University Press, Cambridge (first published 1944).

Wolpert, Lewis (1991). *The triumph of the embryo*. Oxford University Press, Oxford.

GLOSSARY

autocatalysis   The catalysis of a reaction by one of the products of the reaction; can lead to an increase in the concentration of the reacting substances.

catalysis   The acceleration of a chemical reaction by a substance (catalyst), which is unchanged after the reaction.

chloroplast   An intracellular organelle within which photosynthesis takes place.

chromosome   A string of genes linked end to end.

crossing over   The exchange of genetic material between chromosomes by recombination.

cyanobacteria   A group of bacteria able to photosynthesize; also known informally as blue-green algae.

diploid   An organism with two similar sets of chromosomes.

DNA   A molecule composed of two complementary strands of nucleotides, of four kinds (adenine, A; cytosine, C; guanine, G; and thymine, T); the physical carrier of genetic information.

enzyme   A protein acting as a catalyst.

eukaryote   An organism whose cells have a nucleus, containing chromosomes.

gamete   A cell with one set of chromosomes that fuses with another gamete to form a diploid zygote, from which a new individual develops. In animals, the gametes are the spermatozoon and the egg.

gene   A length of DNA, usually coding for a single protein.

genome   The complete set of chromosomes of an individual.

germ line   A lineage of cells from which the gametes, egg or sperm, are produced.

haploid   An organism or cell with one set of chromosomes.

heredity   The tendency of like to beget like. In a system of limited heredity, only a few types can reproduce their kind; with unlimited heredity, an indefinitely large number of different types can reproduce their kind.

hermaphrodite   An individual producing both eggs and sperm.

isogamy   The fusion of gametes similar in size and form; anisogamy is the fusion of dissimilar gametes.

linkage   Two genes are linked if they are on the same chromosome.

meiosis   A process of nuclear division by which a cell with two sets of chromosomes (diploid) produces haploid gametes.

metabolism   The chemical reactions in living organisms, usually involving an exchange of materials with the environment.

mitochondrion   An intracellular organelle in eukaryotes, oxidizing carbon compounds and using energy to synthesize ATP, a molecule used to provide energy for other cellular activities.

mitosis   The typical process of nuclear division in eukaryotes, in which first each chromosome is replicated, and then a complete set of chromosomes is passed to each daughter cell.

mutation   A change in a gene.

organelle   A structure within a eukaryotic cell; examples are mitochondria and chloroplasts.

parthenogenesis   The production of an offspring from an egg, without fertilization by a sperm.

phagocytosis   The uptake of a solid particle by a cell, by budding off a vesicle containing the particle from the cell membrane.

photosynthesis   A process in which the energy of sunlight is trapped by a green pigment, chlorophyll, and used to synthesize sugars from carbon dioxide and water.

plastid   An intracellular organelle containing a pigment; chloroplasts are an important example.

polymer   A large molecule formed by stringing together a series of similar elements, or monomers.

prokaryote   An organism whose cells lack organelles and a nucleus; includes the bacteria and cyanobacteria.

protein   A large molecule formed by a chain, usually folded, of 20 kinds of amino acids. Enzymes, and the main structural components of cells, are made of proteins.

protist   Eukaryotic organisms lacking complex cell differentiation; that is, eukaryotes other than fungi, plants, and animals.

recombination   The physical process whereby the genetic material of different chromosomes is exchanged; it involves the breaking of two chromosomes at corresponding points, and rejoining them after an exchange of partners. The cause of crossing over.

ribosome   An intracellular structure formed of protein and RNA; the site of protein synthesis.

ribozyme A catalyst made of RNA; an RNA enzyme.

RNA A molecule resembling DNA, but single-stranded, and with the nucleotide thymine replaced by uracil. Has several important roles in protein synthesis, and forms the genome of some viruses.

symbiosis The living together of two or more organisms of different species; referred to as parasitism if one organism is living at the expense of another, and as mutualism if both partners benefit from the association.

zygote A cell formed by the union of two gametes.

INDEX

Terms defined in the Glossary are printed in **bold type**

acquired characters, inheritance   10, 28, 140
Altman, S.   39
Alu element   97
amino acids   10
anisogamy   80–1, 92
Anderson, R.M.   106
Antennaria   84
aphasia   150–2
aphids   102
Arabidopsis   118
archaebacteria   62, 75–7
Archaezoa   61, 74
Aristotle   11
ATP   72–3, 77
Australopithecus   141–2
autism   144
autocatalysis   6–8, 12
autotrophy   55

baboons   137
Berwick, R.   157
Bickerton, D.   155, 164
Bloom, P.   165
Boesch, C.   139
Broca's area   150
Butlerov, A.   7

Cambrian   109–10
castes, in insects   128, 132–3
catalysis
Cavalier-Smith, T.   62, 66, 74
Cech, T.   39
cell wall, bacterial   60–3, 77
centromere   68–9
centrosome   68–9
Chaitin, G.J.   15
chemoton,   12, 50
Cheney, D.   143
chimpanzees   139–41, 163
Chlamydomonas   91
chloroplast   25, 70–4, 91, 101
Chomsky, N.   140, 149, 154, 156
Chromista   74

chromosomes   21
   origin of   52–3
   segregation   62, 65–70
cilia   63
Cnemidophorus   79
code, genetic   39–41
   origin of   42–6
codon, stop   40
complexity, measure of   15
conflict, genetic   24, 95–9
Convoluta
creole languages   155
Crick, F.H.C.   39, 172
crossing over   88–90
cultural inheritance   137–40, 145
cyanobacteria   59
cytosis   60
cytoskeleton   63–4

Daly, M.   130
Darwin, C.   1–3, 20, 31, 125–6
Dawkins, R.   139, 153, 172
Descartes, R.   11
DNA
   amoount of in organisms   16
   repair   70, 85, 90
   replication of   8, 10, 33
diploid   79
dolphins   162–3
Drosophila
   antero-posterior gradient   120
   eye development   122
   Hox genes   120–2
   meiotic drive   96
   P-elements   97
Dubos, R.   104
Dunbar, R.   143
duplication, of genes   26–7
de Duve, C.   172
Dyson, F.   12, 172

ediacaran   110
Eigen, M.   35

Einstein, A. 32
endomembranes 64–6
endoplasmic reticulum 60, 65
endomitosis 87, 89
enzyme 5
epigenesis 26, 28
error correction 36
error threshold 34–6
Escherichia coli 112, 120
eukaryotes, origin of 59–78
eusociality 126–8

Faraday, M. 9
fermentation 72
flower development 118–19
Foley, R. 140–1
Fox, S. 32
fungi 103, 106

Gaia hypothesis 48
Gamble, C. 142
gamete 79, 88
Gánti, T. 12–13, 172
gene 17
gene regulation 110–13
genetic algorithm 123–4
genetic assimilation 165
genome 16
germ line 111, 113–14
gibbons 163
Gopnik, M. 158–60
Gordon, D. 133
gorilla 140–41
grammar
  action 166–8
  generative 149, 154–8
Greenfield, S. 166–8
gymnosperms 25

Haig, D. 98
Haldane, J.B.S. 31, 125, 129
Hamilton, W.D. 22, 92, 125, 129
haploid 79
haploid–diploid cycle 87–90
Harvey, W. 11
heredity 1, 76
  cellular 113–14
  cultural 137–40, 145
  dual 113
  limited and unlimited 7–9, 73, 168–9
  membrane 73–4
  modular 8
hermaphoridite, male sterility in 23–4
heterotrophy 55
Hickey, D. 86
Hinton, G.E. 165

Homo erectus 142–2, 144
Homo habilis 141–2
homologous base pairing 8, 39
Hurst, L.D. 77, 92
hybrid vigour 89
hydrogenosome 76–7
hypercycle 48–9

imprinting, of genes 84
inclusive fitness 129
individual recognition 137
information 9–11
intragenomic conflict 95–8
ions 32
isogamy 80–1, 92

Jablonka, E. 29
Jackendoff, R. 172
Jacob, F. 112

Kant, I. 146
Keller, L. 127–8
kin bonding 140–1
kin recognition 131
kin selection 129

language 139, 145–6
  acquisition 152–4
  and future 168–70
  genetics of 158–61
  origin of 149–70
Leibniz, G. 12
Leigh, E. 23–4, 96
lichens 24, 103, 106
linkage 52–3
Luisi, P.L. 57
lysosome 61, 65

Margulis, L. 60, 107
mating types 91–2
May, R.M. 106
meiosis 81, 87–90
meiotic drive 96
membranes, origin of 53–7
Mendel's laws 21
meme 139–40
metabolism 3, 5, 11
methanogens 75
methylation 113–14
microtubules 62–3, 68–9
Miller, S. 31, 42, 53, 55
mind, theory of 143, 146
Mithen, S. 143–5, 165

mitochondrion 22–4, 59–60, 70–4, 77, 91, 102
mitosis 62, 66–7
modules
  in the brain 143–5
  in development 124
  in heredity 8
Monod, J. 112–13, 172
Muller, H. 11
Müller, M. 75–6
Muller's ratchet 83
mutation 1, 36, 45, 83–4, 89
mutualism 101–7
myth 147–8

Nanomia 134–5
natural selection 3–5, 107, 146
neanderthals 142, 162
nepotism 131
Nilsson, D. 153
nitrogen fixation 102
Nowlan, S.J. 165
nucleus 60, 65–6, 70–71

observational learning 138–9
Oparin, A.I. 31
organelle 59, 62, 70–74
orang-utan 140–41
Orgel, L. 8, 35, 39

Paley, W. 153
Paranthropus 142
parasitism 101, 103–6
parental care 80
parent–offspring conflict 97–8
parthenogenesis 25, 79, 84
pattern formation 115–18
P-elements 97, 99
Pelger, S. 153
peptide bond 40
Perrin, N. 127
phagocytosis 63–4, 74, 77
phonemes 168–9
photosynthesis 55, 59, 72
phylotype 122–4
pidgin 155, 161
Pinker, S. 150, 165, 172
plasmid 86
plastid 59–60
pleuromitosis 68
polymer 32
pre-eclampsia 98
Premack, D. 163
Popper, K. 32
prokaryote 59

protein 10, 32
  synthesis 39–41
protist 61
protolanguage 161, 164

Radman, M. 148
rats, learning in 138
recombination 70, 83, 90
Reeve, K. 128
relatedness 21–2, 126, 145
religion 147
Rhizobium 102, 107
Riftia 102, 105
ribosome 40–41, 43–4, 65
ribozyme 39, 42–4
ritual 147–8
RNA 17
  evolution, in test tube 33–4
  as enzyme 36, 39
  messenger 39–40, 66
  transfer 40–41, 44
  world 37–9
Rose, M. 86
rotifers 84

Salmonella 104
Saussurian signs 161–2
Schrödinger, E. 12, 172
Schuster, P. 35, 48
selection
  group 20, 81–4, 147
  levels of 19–20, 81–6
  natural 3–5, 107, 146
self-organization 115–17
Seyfarth, R. 143
sex
  advantage of 81–6
  origin of 79–83
  twofold cost 80, 83
sexual differentiation 91–3
Shaw, B. 83
Sherman, P. 127
signal peptides 66, 73
sign language 155, 169
skew, reproductive 126–8
social contract game 146–8
species 79
stochastic corrector model 50–52
Stringer, C. 142
superorganism 126
symbiosis 25–8, 60, 101–7
  and origin of organelles 70–74
synergy 22–3, 51, 83

template reproduction 1–2, 10, 115
Thymus 23

transcription 66
translation, DNA–protein 39–42
transposon 86, 89, 96–7
Trivers, R. 97
Tyler, L. 152

Újhelyi, M. 163
Urey, H. 31

vervet monkey 143, 161
viruses 12, 42, 99
 HIV 104
 myxoma 105–6
Volvox 92, 109

von Frisch, K. 133
von Kiedrowski, K. 33

Wächtershäuser, A. 32–3
Watson, D.M.S. 15
Weismann, A. 28, 110–11, 114, 119
Wernicke's area 150
White, H.B. 43
Williams, G.C. 84
Wilson, M. 130
Wilson, W.M. 151
Woese, C. 39
Wolpert, L. 119–20, 172
Wynne-Edwards, V.C. 20

zootype 121, 124